Winstanley's Wonders

By
Ian R Farr

MAPLE
PUBLISHERS

Winstanley's Wonders

Author: Ian R Farr

Copyright © Ian R Farr (2025)

The right of Ian R Farr to be identified as author of this work has been asserted by the author in accordance with section 77 and 78 of the Copyright, Designs and Patents Act 1988.

First Published in 2025

ISBN 978-1-83538-824-2 (Paperback)
    978-1-83538-825-9 (Hardback)
    978-1-83538-826-6 (E-Book)

Book cover design and Book layout by:
    White Magic Studios
    www.whitemagicstudios.co.uk

Published by:
    Maple Publishers
    Fairbourne Drive, Atterbury,
    Milton Keynes,
    MK10 9RG, UK
    www.maplepublishers.com

A CIP catalogue record for this title is available from the British Library.

All rights reserved. No part of this book may be reproduced or translated by any form or by any means, electronic or mechanical, including photocopying, recording or by any information storage and retrieval system without written permission from the author.

The book is a work of fiction. Unless otherwise indicated, all the names, characters, places and incidents are either the product of the author's imagination or used in a fictitious manner. Any resemblance to actual people living or dead, events or locales is entirely coincidental, and the Publisher hereby disclaims any responsibility for them.

CONTENTS

PROLOGUE 6

Chapter 1 – THE BEGINNING OF THE END 7

Chapter 2 – A LOAD OF BALLS 21

Chapter 3 – THE ROOM OF WONDERS 26

Chapter 4 – AN AFFAIR OF THE HEART 30

Chapter 5 – LONG LIVE THE KING 34

Chapter 6 – EXPLORING A NEW WORLD 39

Chapter 7 – A GOLDEN OPPORTUNITY 45

Chapter 8 – THE CONQUEST 50

Chapter 9 – A CLERK AMONGST CLERKS 57

Chapter 10 – SEIZING OPPORTUNITIES 59

Chapter 11 – OH, FATHER, LEAD HIM INTO TEMPTATION 64

Chapter 12 – NATURE DECIDES 67

Chapter 13 – WHAT'S GOOD FOR THE GANDER IS NOT
ALWAYS GOOD FOR THE GOOSE 77

Chapter 14 – A CHANGE OF FORTUNE 82

Chapter 15 – DESPAIR 89

Chapter 16 – THE CRUEL SEA 98

Chapter 17 – FRANCE 102

Chapter 18 – LETTERS OF LOVE 106

Chapter 19 – MEANWHILE… 111

Chapter 20 – THE PALACE 113

Chapter 21 – CONSEQUENCES 122

Chapter 22 – HEADING HOME 125
Chapter 23 – LED BY GREED 137
Chapter 24 – LOVE VERSUS AMBITION 150
Chapter 25 – TRANSFORMATION 159
Chapter 26 – ANCHORS AWEIGH! 162
Chapter 27 – ACCELERATION 169
Chapter 28 – THE HOUSE OF WONDERS 173
Chapter 29 – PREPARATIONS 175
Chapter 30 – THE GIFT OF INDEPENDENCE 178
Chapter 31 – ABSENCE 188
Chapter 32 – LONDON 191
Chapter 33 – CHANGES 197
Chapter 34 – AN UNREAL REALITY 200
Chapter 35 – A TIMELY MEETING 211
Chapter 36 – AT ODDS WITH HIS REPUTATION 214
Chapter 37 – OF PELICANS AND RECONCILIATION 218
Chapter 38 – A FOOLISH FOLLY 224
Chapter 39 – A TRAGIC LOSS 226
Chapter 40 – OUT WITH THE OLD AND IN WITH THE NEW 235
Chapter 41 – OPENING NIGHT 237
Chapter 42 – DEVIOSITY 245
Chapter 43 – LIFE'S TRUE PURPOSE 255
Chapter 44 – A MEETING WITH THE DEVIL 259
Chapter 45 – PEPYS V. WREN 266

Chapter 46 – THE BEGINNING OF THE END - COMPLETE 273

Chapter 47 – REPETITION 280

Chapter 48 – THE COURAGE OF IGNORANCE 291

Chapter 49 – BY ROYAL COMMAND 294

Chapter 50 – THE GODS ARE ANGRY 298

Chapter 51 – JACK AND JANE TWO DAYS EARLIER...................... 302

Chapter 52 – THE CITY OF SHIPS.. 306

Chapter 53 – HENRY'S BABY 316

Chapter 54 – THE WRATH OF GOD 327

Chapter 55 – THE END OF THE END 330

EPILOGUE 332

PHOTOGRAPHS 334

FOOTNOTES... 343

ACKNOWLEDGEMENTS 344

PROLOGUE

Dear Reader,

You may never have heard of Henry Winstanley Gent., wonder why he is called 'Gent.', and indeed why he is worthy of a novel at all, but 360 years ago Henry was very famous indeed, embodying the perfect combination of an imaginative mind born into an age of invention where discoveries that would one day be taught in schools, were germinating in the minds of their creators.

An Inventor, an Engraver, a Showman - and many would say a Madman – Henry could list Kings, Princes, the world's most famous Diarist and the greatest Scientific mind of his time as his admirers.

Many found his creations too whimsical, but Henry achieved something others greater than himself, believed impossible...

CHAPTER 1
THE BEGINNING OF THE END

Our story begins on the 25*th* day of the month of June in the year 1697 at 5:45am or thereabouts, approximately 9 miles Southwest of the coast of England.

Under a baby blue sky devoid of cloud, a band of 8 foolhardy Britons were risking their lives on an ugly chunk of rock to create something rather wonderful born from the mind of Mr Henry Winstanley Gent...

"Oh, my Lord - look to the horizon, Henry."

"I always look to horizons, Tom."

Henry span on his heels following the line of Tom's finger pointing into the sun and gasped, "The sun is toying with me - warming my cheek yet blinding my eye."

"The enemy is but ten minutes away, sir."

"Confound it, where is 'The Terrible' when I need her?" Henry roared, panning his 'scope to left and right.

"I saw her heading South by South-East at full sail some time ago. We are doomed, sir."

"Don't spout such gibberish, man - we are not doomed until I say so."

Thrusting a hand into his pocket Henry's fingers eventually found what they sought - a slice of blue stained glass, which he placed on the end of his 'scope, and for the first time was able to see what Tom had been referring to – an enemy Privateer under the French flag slicing through the waves to reach him, "We are doomed, abandon rock."

Yet Henry's men remained steadfast, showing their loyalty to this eccentric man of advancing years in his noble cause - rowing 9 miles there and 9 miles back almost every day for weeks on end, in a small boat to strike dents in a lump of rock in order to protect the Men of the Sea from one of its monsters - the mouth of the Eddystone and its teeth of rock -

Madness.

"Save yourselves men – I'll wager young Captain Bridge has abandoned his post in pursuit of glory."

But Henry's men still chose to disobey him, standing resolute on that chunk of rock guarding the minor miracle they had been creating under his guidance – from the spark of an idea 2 years ago to the first spark of his pick two months later. From that day forth many more sparks had flared as his crew set about denting the rock in order to plant the twelve iron fingers that now pointed to the sky like in a giant's hand. Cradled in the palm of that hand was Henry's baby, growing into the design he had first committed to paper 24 months and a lifetime ago between the flashes of a violent thunderstorm in his garden studio at Littlebury, 44 miles North of London in the bountiful county of Essex.

Henry was confident it would have eventually met with his father's approval – a difficult if not impossible task, for his father had always dismissed his son as a whimsical fool – words that still hurt 40 years later.

The French ship finally arrived kissing the Rock, decanting ten of its countrymen to negotiate the swell. Henry's sea-hardened crew grinned as they watched the portly French Captain stumble with knees bent at an angle not intended in God's design, sending him crashing belly down onto the rock, adding several ribbons of kelp to his braids of gold.

The Captain's first word confirmed both his nationality and his mood - "Merde!"

Henry, however, vowed to maintain the manners instilled in him since birth, as an Englishman of good character,

"I beg of you, stop, sir - you are wrecking my life's work."

The Captain chose to reply in English, "Your work here is over as soon will be your life," then addressing his men, "Debarrasez-vous d'eux" (Get rid of them).

The French sailors took great delight in stripping Henry's men of their clothes, readying them for a gruelling trip home rowing headlong into a brutal westerly with no clothes on their back, and nothing to give their wives for the family table. Yet even so they stayed defiant calling,

"God be with you, Henry."

Failing to have earned the same respect from his own crew, the French Captain sneered with envy, leading Henry to conclude that the character of a man can be gauged by his smile – an open smile being produced by an open mind, whereas a mean man could only ever conjure a sneer. Henry was convinced that if he had a pen and paper to hand, he could have sketched an accurate likeness of this man's mouth, even when blindfolded - which was useful considering what was to happen next…

The Captain's sneer, along with the face creating it, suddenly disappeared at precisely the same moment that Henry felt a blindfold being tugged across his eyes. Deprived of sight Henry's hearing grew stronger to compensate - the shuffling of shoes, the slapping of waves, the cries and the occasional thud which could only mean one thing: his crew were being pushed into the sea.

"I beg of you - do not harm my men."

The sudden blast of cold against his chest indicated his tunic had been ripped open. Then a voice hissed in his ear, a sneering voice within splatterings of saliva:

"Je m'appelle Capitaine Lazelle. Je suis heureux de faire votre connaisance, Monsieur Winstanley, surtout dans mes conditions." (My name is Captain Lazelle, I am happy to make your acquaintance Mr Winstanley, especially on my terms).

Henry struggled to reply through chattering teeth, "Pray inform me of your intention, sir."

But as soon as the words left his mouth Henry knew they were wasted for the sudden blast against his buttocks meant his leggings had been yanked down, the pain shooting up from his knees meant he had been pushed down to kneel on the rock and the shock of spray across his back confirmed his tunic had been completely torn off.

"There is no need for this, Captain – how can I pose a threat to France with my pants down?"

The Captain answered in English, "'I have been warned not to underestimate you."

"That is good advice sir – but you have me at a disadvantage – who flatters me thus?"

"You will discover soon enough."

Henry cried out – ribs of metal were tightening around his chest signifying a chain was being coiled around his body like a snake around its prey. Captain Lazelle spoke again,

"Listen to me, Henri and listen hard."

"You have me there, sir, I cannot do much else."

"Emmenez-le chez Le Roi"

Henry had heard those words before and they did not bode well,

"Le Roi monsieur?"

"The King."

The explosion of laughter caught Henry by surprise, not from any French accent for that would have been expected, but from the sea birds mocking from above with their infernal cackling, and they remained cackling as he was tugged along a plank, heaved over a hatch and dropped into the bottom of the ship's hull where he collapsed in a heap, hearing his captors laughing above as they set about demolishing his work.

Maybe it was due to the pain in his knees or being blind, naked and freezing cold, but Henry fancied he was starting to hallucinate. The hammering from above deck didn't help matters, but neither did the movements all around him in the bowels of that ship. Something was shuffling in the shadows encircling him, which he guessed to be an army of small creatures - and they were closing in.

Henry knew not to panic - his best defence being his wits and strength of character, both of which he held in abundance.

The shuffling was joined by a new sound – grunting. Despite himself Henry was beginning to shake, kicking out at the creatures encircling him until he realised the noise was coming from inside of himself. Whatever it was, it had taken control of his body, making him convulse as it raced to escape, surging up his gullet like a torrent of molten lava forcing his jaw wide open to decorate the floor.

Not a man to normally suffer sea-sickness Henry concluded that as things were far from normal there was no other conclusion - for the first time in his life Henry Winstanley was being sea-sick.

Left alone in complete darkness Henry guessed the enemy's intention was to make him feed on his own fears until his mind turned against itself, so decided to be calm and concentrate instead on the things he knew best, and the thing he knew better than anything else on Earth was the man called Henry Winstanley Gent, through the experiences that had turned Henry the child into Henry the man. And so, it was there, in that dark hold of the enemy ship that the chance came for reflection. And the more he reflected the faster his memories flooded in.

The earliest memory transported Henry back to the smells of his childhood - for smell is the most evocative of senses - and in particular his mother's cakes on the occasion of his breeching party. He had turned 5 years old, surrounded by his family, parading around the house in his new pair of breeches. His two brothers, his sister and his parents were all present in this perfect recollection, and all of them smiling, apart from his mother. He could not have understood at the time, but later realised she was grieving the inevitable separation of a mother and son. She had tried her best to adopt a brave face but losing a son to manhood was a grave loss and his breeches had made it all the more obvious - his sister Susannah had remained in a skirt, jealously watching both of her brothers jumping over gates and onto ponies with their dignities intact.

As long as she remained in a skirt that freedom would be denied her.

He remembered how sad his mother had looked that day knowing the time would eventually come when her sons would ride away as young men never to return as boys again. At least she knew that Susannah, constrained by her skirt, would remain in the family fold far longer, possibly even for the rest of her life…

…meanwhile, back in the French ship an idea came to Henry – as he was already on his knees, why not attempt a prayer? Harbouring no belief in the Almighty, Henry thought he may as well hedge his bets. He smiled, solely for his own benefit as there was nobody else nearby to see, but it felt ironic that the father who had failed to give him guidance throughout his childhood, was the figure he was turning to now for that very same thing.

Henry clearly remembered his father as Church Warden speaking kind words in the House of God yet never sharing such kindness in his own.

In his frustration Henry had dismissed Religion and began to search for another guiding light - and in that search he had discovered Science, which satisfied his curiosity and soothed his anger. Coincidentally, at the same time his brother Robert came to the end of his own search - making good use of his anger by influencing others through the ink that flowed from his quill onto leaflets that others could read. Robert's words however, ceased to flow once the Plague took hold, rendering his fingers unable to control the nib, and showing no mercy, the Reaper attacked swiftly from behind and plucked the sparkle from his eyes.

The Reaper consigned Robert to the communal pit that had been dug by young men whose strength would have been better served planting seeds for the community rather than burying them.

The death of his young brother had left Henry to cope with Life on his own terms. His mother had deserted him when he needed her the most, shutting herself off to cry while his father slid away to the sanctuary of his church. In his desperation Henry also entered the church, quite literally, raced past his father, took a bottle of the Communion wine, rushed outside to his brother's grave, smashed the bottle on his gravestone, collapsed to his knees and called down the makeshift megaphone,

"God bless you, Robert. I miss yo -"

No sooner had he started than Henry was tugged off his feet by his father,

"How dare you use the Lord's name in vain!"

"I just wanted to say goodbye, father."

"Then go inside and say it in prayer."

"Forgive me, father, but he has a better chance of hearing me this way."

He may not have heard his son or chose not to argue a moot point but either way Henry's father spoke no more, and Henry did as he was told, carrying his burden inside alone.

Over the following days Henry's father had attempted to console what was left of his family, explaining that the Lord in his kindness had transported Robert to a better place. Yet his mother had nothing so glib to hide behind, and suffered terribly as a result. As his father rang the church bell day and night, it seemed to Henry that the bell's toll worked as a pump draining his mother's cheeks of blood with every chime.

His other brother Charles – such a robust youth – had so far denied the Reaper yet teased Death by joining Cromwell's forces in that dreadful Uncivil War. He remembered his mother watching her son for the last time, joining a band of boys riding off into the sunset looking identical in their uniforms chattering like a flock of birds disappearing beyond the fields of rye for evermore…

His mother was left to console herself, that at least Charles had looked happy the last time she had seen him.

And so, the smiles that had once graced the family table grew fewer over time until only Henry's remained. His mother stayed at home crying until her ducts ran dry, whilst her husband chose the Lord's House to disappear in prayer - but to Henry's mind dispensing words from the mouth was far easier than leaking tears from the heart.

Henry remembered hearing his father extolling the same message week by week, month by month, to a grateful congregation that grew smaller every Sunday, and as he lay to sleep each night Henry wondered if his father was actually gaining pleasure from becoming the most influential member of that community...

...back in the belly of the French privateer Henry woke to something nibbling at his toes, doubtless making a meal of the vomit he had provided. Companionship was a much welcome friend and if it had to be from one of the lowest forms of life, then so be it, for lying in his coffin at the end of days Henry knew he would be completely at their mercy - the tiny creatures who cared not a hoot whether his allegiance be with England or France, if there was meat in the offing. In fact, Henry concluded that in many ways he was being treated fairer by the rats than he was by the French.

Thud, thud, thud...

His feet suddenly jumped from the ship's hull beating against the waves below, feeling it sway under the sea's demands leading Henry to conclude they must be turning into the wind, hitting the waves head on.

After a few minutes the thuds grew weaker, until Henry fancied, he could feel the ship lift then glide across the surface as if entering a place of shelter - his final port of call.

A voice in his ear carrying the stench of garlic, grunted words he did not understand. Hands clasped his arms tight, and Henry felt himself being lifted to his feet and harried up the steps from the bowels of that ship onto its deck.

The air was fresh and a welcome change as Henry readied himself for the grand coach that would transport him to his judge and executioner – Le Roi, Louis X1V, The Sun King. He was expecting a gold coach at the very least...

But it was only the cackling gulls that greeted him along with a familiar but dreadful sound – the clanking of chains and the thuds of something obviously heavy by the tone of the voices dragging it.

The blindfold was tugged away but he faced no decorative coach, just a plain stone wall – a featureless ugly wall, and the mouth of Captain Lazelle contorting into a sneer once again, emitting the same vile stench as before when he spoke,

"Welcome to your final resting place, Henri."

Henry fell to his knees guided by a brutal force - a boot applied to the back of his knees sending him tumbling down the next flight of steps – no longer wooden but stone. He felt the temperature drop as he descended into the tiny cube of a cell in which to spend his final hours on Earth, chained to a French cannonball, the weight of which he had heard them cursing only seconds before.

Left alone again Henry tried to dream up a device to set himself free, but even he struggled to invent anything at all with the meagre resources at his disposal, so resigned himself to sit and think - always a useful exercise to his mind.

The scientist in him believed that every problem has a solution, and the best solution was rarely the obvious one: a memory flashed into his mind as vivid as if he was standing there once again, amongst the crowd of onlookers in the East End London Docks watching a muscular little man, chained like Henry, taking three deep breaths before being plunged into a barrel of water.

He remembered noticing how the water barely moved after the bubbles had vanished and despite realising this was purely a stunt, Henry could feel his heart thumping against his chest so loud he thought it would disturb the others. Yet the crowd's attention was fixed firmly upon that barrel, and everyone's heart was no doubt racing just as hard as his.

A cry went up as bubbles suddenly frothed over the barrel's rim and, despite himself, Henry had moved forward to help. The crowd gasped, a man cried, and a great whooshing brought an explosion of applause; the muscular little man was pulled from the barrel, and proceeded to bow in all directions to the elated crowd. Everyone present did not begrudge the penny they had spent to witness such a thrilling event, and as the crowd dispersed Henry was the only one who paused to think - this muscular little man had obviously survived the stunt many times before so he put his mind to work to find the trick – for being a showman of sorts, Henry was convinced there had to be a trick. Eventually it came to him as he watched the man walking away, for Henry noticed he looked less muscular than before leading him to conclude the man must have expanded his muscles before the chains were tightened then contracted them at the end of the show allowing the chains to slacken and fall.

So that was it then - logic, experience and bravery had combined to save this showman's life and provide supper for his family's table. Henry knew he had to apply those same attributes here to save his own life, which he listed to make his plan clear:

Logic – he had been captured by the enemy after being deserted by an Admiralty ship tasked with his safety.

Experience – not good - he had no experience of prison.

Bravery and Skill – he had proof of those in abundance.

Skill – his skills needed pen and paper with which to design, plus materials with which to build - both sadly lacking in that miserable place.

Yet as he scanned the cell his eyes settled on that large cannonball, and it brought back more childhood memories - of a straight line he had drawn on a ball that appeared curved when viewed from the side. He remembered seeing that same curve as the join on his brother's helmet allowing it to fit over the dome of his head, and then, whilst standing in the garden, he had noticed a narrow cloud that had traced the same curve in the sky above his head from baby blue directly overhead to French blue, by the time it touched the horizon.

So that was it then - Henry concluded he was living under a dome.

And suddenly it all made sense - Senor Galileo was right - the Earth was a giant ball as was the Sun.

Henry suspected his conclusion would not sit well with his parents, and yet how could they disapprove of something that was so obviously true?

At least his mother might see reason…

"Go up to your room now and do not come down until you have denounced everything you just said. Your father will be furious. Stop being ridiculous – the world is flat, or we would all fall off," she had said.

"If that is all you have to say on the matter, I will go and ask father."

He banged open the back door at precisely the same moment as a massive thunderclap exploded in the heavens above, resounding across all the neighbouring gardens setting their dogs a-barking.

Henry was determined to meet his father face to face so despite the impending storm, sprinted for the church hearing his mother above the thunder, screaming from the house behind,

"Come back, Henry, I do not wish to lose you too!"

But she was too late – Henry was already sprinting away down the garden, curious as to why, despite being only two o'clock in the afternoon, the sky was as dark as night, as if the Sun had deserted it.

Searching the heavens but finding only clouds, Henry watched wide-eyed as those same clouds were caught in an orbital trap like water spinning down a hole. He barely recognised their garden anymore – a swirling mass of branches circled the air being expelled in all directions, smashing remnants of his childhood to tatters - from the swing on which he had first experienced how energy can turn from potential into kinetic, to the bridge from where he had watched tadpoles sprout legs, and transform into a completely different species.

The wind screeched as wild as a banshee as did the metal weathervane atop the church as it span in demented fury. Women screeched too, caught by the gust, trying to calm unruly skirts from exposing their buttocks, whilst men cursed it for exposing their bald scalps. Thus both buttocks and scalps raced together for the shelter of the church led by Henry's father calling,

"Go inside Henry, the Lord will protect you."

Henry raced inside witnessing the townsfolk pledging everything they owned - their chickens, pigs, wives and daughters to the Lord in return for His protection. Upon seeing his son, Henry's father could not contain his joy calling to the Vicar,

"Look Father, my son is finally seeking our protection."

But the Vicar's smile proved premature as Henry growled,

"Not protection father; I've come to witness the Lord's mischief."

Henry's father cried out loud so all could hear, "This is not the Lord's mischief, but that of the Devil!"

A brilliant flash followed by a devilish bang as loud as any cannon, sent the congregation crashing to their knees including the Vicar and Henry's father, acknowledging this greatest Power.

Of all the heads in that church Henry's was the only one to look up and therefore was the only one to witness the Sun returning Light to the Earth.

And as the light returned, it exposed the God-fearing townsfolk down below trembling under the pews – yet Henry's attention remained elsewhere, fixed on the sky behind the stained-glass windows, trying to make sense of what had undoubtedly been a historic occurrence, beyond both his comprehension and the control of Man.

As he stared at the heavens, Henry saw a massive shape akin to a ball slowly slide across the face of the Sun, its rays dimming before re-emerging as spears of light. A deafening blast released a lightning bolt at 270,000 mph heading straight for him. Yet Henry had detected another sound – a distinct 'fizz', that he reasoned to be somehow connected to the flash. Through the stained glass he saw the tree he used to climb as a boy glow red for an instant before falling towards him, its branches held out like arms in self-defence punching through the stained-glass window sending shards of the Virgin Mary raining down around him. Holding one up to his eye Henry gasped – protected by the blue stained glass he could now clearly see the huge ball sliding across the Sun – a ball that he reasoned to be the Moon.

Young Henry cried out, elated at his discovery but scaring the congregation as a result. He sprinted for the door, desperate to reach home and find the one thing that could explain everything.

Racing through the kitchen he met his mother, shocked but delighted that he was still alive, trying to hold him tight, blocking his way. But Henry had no time for any of that,

"Where does father keep the Works of the Devil?"

Caught on the hop, his mother hadn't the wherewithal to delay him,

"The Devil's Works?"

Henry gripped her hand tight, "Yes mother, I must destroy all traces of Satan in this house."

Her reply was both eager and swift, "Oh Henry, your father will be so pleased – they are upstairs in his study."

Freeing himself from her grasp Henry raced for the stairs sending his mother into a delightful spin, as he bounded to the topmost step praying his father hadn't burnt those 'Devil's Works' just yet, for Henry was convinced their worth was far greater than any scripture his father had preached in that church.

Reaching the top step Henry steadied himself, gasping for breath from being within touching distance of everything his father had kept hidden from him - here and now shielded by nought but a crude sackcloth curtain mere inches from his face.

Tearing it aside, Henry faced a pile of books his father had stacked in readiness to be burnt – each one plump with pages deemed to be heretic by those who felt the most threatened by them – the teachers and preachers of the land.

His eyes opened wide absorbing the words now so clear in front of them - of Senor Galileo's Discoveries made from staring at the heavens through a telescope, that now threatened the teachings of those who had preached of the Earth's superiority in the Cosmos, being postulated here as only a part.

Henry came to a conclusion: – from the orb called the Earth that sails around the sphere called the Sun, to the ball of an Apple that had hit the ball of the eye of Mr Newton that led to the Law of Gravity that affects everything on the planet -

Life was clearly just a load of balls.

CHAPTER 2
A LOAD OF BALLS

There was little for Henry to see in that gloomy cell apart from the straw at his feet and the huge cannonball serving as his companion.

So instead, Henry chose to look inside of himself and in so doing remembered another time when his life had been threatened: he was only 24 at the time but the memory and the words were still vivid...

"Hurry Henry, or the Earl will have your head!" ...

Thus, the present-day Henry flinched at the memory of his earlier self flinching upon hearing his friend Jack calling from three floors below.

Henry was in his favourite place on Earth – an isolated room behind the eaves of Audley End Manor House and did not wish to be disturbed. High up in glorious isolation Henry could experiment as he wished alongside a window to the heavens.

Jack's voice pressed home his point, "Henry, can you hear me?"

But Henry could not, or at least his mind was so preoccupied he blotted out all distractions. A globule of his spit hit the surface of a huge brass sphere before being buffed by Henry's cloth as he lovingly polished it into oblivion. Every day he would spend hours in that room experimenting, inventing and building, safe from visitors and their prying eyes, but today was no ordinary day; witnessing his reflection growing clearer as he polished the surface, Henry spotted a sudden rectangle of light indicating the door was being thrown open, revealing Jack's reflection racing up to a halt,

"My word Henry, what in heaven's name is that?"

"Congratulations Jack - Heaven is exactly the right word."

Jack stumbled, dazed and confused, having no idea what Henry's strange contraption was, but as Henry had designed it Jack knew it must be special.

Henry set about clarifying it all,

"Each of these balls is a planet, sailing high in the sky for all who have the will to see, through their windows and over every roof in the land."

Jack stood stock still, eyes and mouth wide open,

"They look like juggling balls."

"Ha, absolutely right Jack, balls of all sizes spinning in the air together, yet never touching."

"Your head will spin off your neck if you don't go downstairs now Henry - the Earl is entertaining an important guest."

"How important?"

"Important enough to have brought his own musician."

"I see, then I shall depart forthwith."

Powered by the exuberance of youth, Henry rushed to the kettle which was bursting into tune from the steam forcing its way through the blowhole of a miniature pewter whale fixed into its spout. Popping it out Henry poured the steaming water into an ornate pot, placed a silver platter onto a wheeled frame and hopped on board. Tugging an overhead chord made the stairs revolve to form a ramp allowing young Henry a swift exit on his adapted skateboard, banging open giant doors as he motored along corridor after corridor beneath the noses of previous owners long since dead, trapped scowling from their frames.

Speeding along the entire length of the Great Hall, Henry realised he hadn't yet tested the brakes. A door with a particularly ornate lock was approaching fast, threatening to leave its pattern etched into his cheek. Henry did not pray much in his life, but this was an exception,

"If there is a God please help me now," and as if on command the door opened, just in time for Henry to speed past the bewildered servant, into a room of serenity wherein a harpist was playing a delightful tune to entertain Thomas Howard, the Third Earl of Suffolk, and his mysterious guest who was gazing out at the gardens beyond.

Grateful for the huge rug that slowed him to as graceful a halt as the music, Henry delivered the fashionable contents of his pot to its intended recipients. He stared, captivated by this guest's hair which cascaded over the back of his chair like a huge bunch of black grapes fresh from Covent Garden Market. At his side the Earl's jowl quivered with every word of his sales promotion,

"I trust you find my House agreeable; I have endeavoured to maintain its excellence at great pains to myself and my purse. It enjoys the most beautiful views over many miles in every direction, and supports the very fine River Cam …"

For a while, as the guest remained listening politely, Henry took the opportunity to check the contents of his box for breakages, sighed with relief at their condition, then extracted a pair of delicate China Cups disturbing the Earl's sales pitch. Earl Howard grunted, Henry flinched, but as his boss turned to face him Henry detected signs of relief at the interruption.

A nod to Henry set into motion an elaborate performance of a simple task: the popping of the ornate whale from the spout, the covering of its blowhole followed by the opening of its jaw allowing the flow of the steaming brown liquid from the belly of the whale into both China Cups beneath.

Henry served the guest first as the Earl continued his platitudes:

"Please accept this drink as a token of my gratitude for your most generous visit - the latest beverage brought from The East on one of my ships – it is called Tea…"

And as the guest turned, "…your Majesty."

Henry instantly flushed with excitement – so this was the new King whose name was on everyone's lips, now less than five feet from his face.

Henry proffered the cup nervously, and received a smile in return. His Majesty did not look the least bit imposing, producing words that were tinged with humour and betraying a hint of French:

"Spurr!"

His Majesty spluttered the drink over his clean white shirt,

"A tad on the hot side Earl, I am used to it being served somewhat cooler."

Henry was aghast, believing himself to being instantly relegated to the ranks of the unemployed, yet upon taking another sip the King broke into a smile as he stared at his own likeness painted on the bottom of the cup – revealed by the emptying liquid.

Never one to pass over an opportunity for self-promotion Henry bowed with appropriate depth,

"I am aware the likeness does not do justice to Your distinguished features, but I hope it pleases Your Majesty nonetheless."

King Charles faced him eye to eye, allowing Henry a privileged view usually reserved for those of Nobility, Members of Parliament, Captains of Industry or pretty Ladies of any social class,

"An agreeable likeness, young sir. I beg, what is your name?"

A delighted Henry bowed as elaborately as holding a pot of steaming tea would allow,

"I am your humble servant Master Henry Winstanley, Your Majesty."

"Well Master Henry, you have earned a penny for your bravery and a shilling for your artistry."

The coins were perfectly aimed, each dropping into Henry's purse with a 'clink' indicating they had joined many more - each coin bearing the likeness of this new King.

Beaming from ear-to-ear Henry bowed out of the room satisfied he was leaving behind a good impression, hopped onto his tray and scooted back along The Great Hall towards his workshop dispensing items he had mended, to their owners along the way - a pair of spectacles for an elderly card player who had lost money for every day he couldn't see his hand, and a necklace for the lady whose hand he was trying to win - each donating their fees to his purse with a satisfying 'clink.'

CHAPTER 3
THE ROOM OF WONDERS

Henry burst into his workshop whistling a gay tune that stopped the instant he caught sight of his father trying to set his contraption into motion.

"Be careful with this one, father, I have yet to test it."

But his father showed scant regard for his son's device prodding and slapping until it sprang into life making him jump in alarm. Henry gazed at his creation and smiled - the planets were performing exactly as planned, each rotating on their metal stems circling the large central sphere of the Sun.

"What do you think to my creation father?"

Mr Winstanley Senior examined it with hands on hips nodding sagely every now and then, "Pure whimsy, perfectly suited to our superficial King. It seems to me a great waste not to apply your undoubted skills to more appropriate ends."

Henry felt dejected, and being young, failed to hide his frustration,

"You have never understood me father!"

"Is there anything *to* understand?"

"Not if you don't look. My device could be taught at the Academies of Science. It shall be hailed as a gift from Mr Winstanley Junior to the nation for the common good of all."

"You are deluded; it fails at being 'good' but succeeds greatly at being 'common'."

Henry was crestfallen but too proud to show it, choosing diversion over confrontation,

"As you are here father, could you advise me on a private matter that troubles me greatly?"

His father's demeanour changed in a flash, having been afforded this elevated position of counsel, "Very well, son, proceed."

"My question is simply this – is it a sin to harbour ambitious desires? I believe something truly great awaits me."

Henry noticed his father's eyes narrow upon realising his son had arrived at the crux of the matter, "Visit the asylums to witness delusion, study History to learn of ambition!"

"I do, father, and History is being made here and now…"

Henry took the opportunity to throw a protective sheet over his creation,

"I have studied Senor Galileo's theory stating it is our world that sails around the Sun and not vice versa." Despite the grunts of a disapproving father Henry was on a roll and continued pressing home his point, "I do so admire Mr Newton for his thoughts on gravity – I even hear that he has unravelled the mysteries of the rainbow!"

His father's temper, having been simmering throughout, had now reached boiling point, "The world of Science is challenging many perceptions but hear me on this – forget your foolish dreams, the only true path is the one that leads to God."

"My dreams rest in Science and Entertainment."

"Ha - Blasphemy and Whimsy," his father groaned, striding out of the workshop leaving his son wondering how to respond without cussing, for Henry had been raised to be a Godly boy.

However, he was no longer a boy – "Wait, God damn you!"

But his father had moved away. Relieved that his father hadn't heard him Henry grabbed a small box and, by revolving the stairs as before, slid down managing to catch his father just as he was entering his private room below,

"Stop and listen to me, father - you always make time for your precious congregations so allow me the same privilege – the son-to-father talk we never have."

His father appeared to be nervous of this proposition and span around in his doorway blocking his son out. But Henry was both angry and fast, lunging forward trapping his shoe in the gap forcing it back open,

"Stop and listen to me father – your time is so taken up with matters of Death I wonder if you are starting to worship it. Your Godly sermons did not protect Robert, and Lord only knows what happened to Charles. I may not have much time left but I assure you I am not going to waste it – I intend to achieve something of consequence. With the ideas inside my head, I can achieve that goal."

His father retreated staggering back to the fireplace,

"You have so much trivia churning in that head of yours there is no room for anything else." His father turned back to face his son clutching a crucifix to his chest as if in defence, and in that moment, for the very first time in his life, Henry noticed fear in his father's eyes as if all his steadfast views were under attack,

"History will remember many things Henry – of how families were severed by that Uncivil War in order to place one silly man upon one fancy chair, of how many brothers who played under the same roof at Christmas would die on opposite sides of the same field by Spring. It pleases me to think you will not die as a man of war Henry, but it hurts me to think you will die as a maker of toys."

As was so often the case, Henry backed away with bowed head,

"I am sorry you feel that way father."

Sensing his son weakening, Henry's father launched his attack,

"Hear me on this - you have ambition, and you do have skill but what you lack is direction! Christopher Wren builds structures with towers as fingers pointing to God! Take your lesson from Mr Wren!"

His words appeared to have the desired effect - Henry turned limp, his spirit defeated,

"I am sorry you think that way father, but I will not view entertainment as a sin."

Henry senior approached Henry junior and laid heavy hands upon his shoulders, presumably intended as an act of care, but it felt more to Henry like an act of control,

"Entertainment is nothing more than fooling around and it makes you nothing more than a fool by doing it. God has given you skills, Henry so use them wisely – aim for the head not merely the funny bone."

"Thank you, father. Some would say I already do…" Henry opened his pouch and, in time-honoured fashion, placed half the contents of his tips into a bowl under the cross.

Henry felt relief as his father's hands lifted from his shoulders, which stopped the instant he saw them joining together in prayer, once again shutting him out.

Placing the box from his workshop onto the mantelpiece, Henry pressed a switch and watched his small creation spring to life - a clockwork motor revealing Christ on a crucifix behind a small central candle. His ingenious mechanism created a spark which successfully lit both the candle and a ray of hope in Henry that his father might offer his approval.

"Pure whimsy, you suit your new King well…"

Left dejected as he had been so many times before, young Henry made a decision – that 'young' Henry was no longer a child, and that 'adolescent' Henry would never buckle under his father's disapproval again.

Placing two fingers upon his wrist Henry detected a strong steady pulse – proof enough that he was still alive and strong enough to face whatever opportunities Fate had in store – be it in the realm of Science or Entertainment, or maybe even Affairs of the Heart…

CHAPTER 4
AN AFFAIR OF THE HEART

To the good folk of Saffron Walden their clock was like a friend - residing in the heart of the place, it provided the town's pulse.

A passing gull on any weekday could spot the townsfolk down below like an army of ants scurrying this way and that on their daily errands – visiting the butcher, the baker and for the wealthier few - the dressmaker.

One such place was 'Taylor's the Tailor Shoppe' - a modest establishment serving as both home and business to the hard-working Taylor family, toiling together on garments of differing quality paid for by differing sized pockets. Their only child was 19-year-old Jane, an attractive girl with a stern countenance, especially now when she harboured troubling thoughts that required airing...

"Father, would you say I am beautiful?"

Her father, a plump fellow with a joyful manner in keeping with his appearance, was concentrating on a particularly intricate stretch of stitches but, never one to deny his daughter attention, glanced up and inadvertently dropped a stitch as a result. He was able to conjure a smile nonetheless,

"My darling Jane, I think we have named you most inappropriately, for you can in no way be considered plain. Your beauty surpasses any other young lady of this fair town – nay, even of the whole county."

Mr Taylor gave his daughter a reassuring hug as he left the room which failed to reassure her at all. Dissatisfied with his sycophantic response, Jane turned to her mother for an honest answer,

"Mother, would *you* say I am beautiful?"

Her mother was a matter-of-fact kind of woman, so answered in a matter-of-fact kind of way, "That would depend upon whom I was saying it to: if it was to Ben, the farmer's son, I should say that when God gave out beauty He thoughtlessly passed you by, and when He dealt out strength to toil in the fields you were equally neglected. To Thomas, the butcher's son, I should say you were on the brink of blooming - but to Jonathan of The Priory I would say your beauty surpasses the glory of Venus."

"I think I understand what you are saying. Thank you, mother."

Mrs Taylor put down the garment she was adjusting and took a moment to regard her daughter, and as she regarded, she realised she had been unkind,

"If you are looking to your parents for guidance, it would be wise to remember your father is as soft as the wadding in his jackets!" to which mother and daughter shared a laugh. "There will be a man for you though. We just need to decide who…"

Young Henry Winstanley had good cause to smile as he strode along the corridors of Audley End Manor House that day towards the Main Room for a meeting with the owner of the house – The Earl of Suffolk.

A particularly grand, and more importantly, wealthy friend of the Earl was approaching him at speed from the opposite direction with hand outstretched, full of coins. Henry's hand was likewise outstretched, gripping his purse plus the wealthy man's repaired spectacles. Thus, Wealth and Youth came together that morning in the manner of a medieval joust: no blood was spilt, the exchange was swift and to each man's benefit – the wealthy man being able to count his wealth more clearly and Henry being able to fund another of his inventions.

Upon entering the Main Hall Henry joined the line of servants who had been assembled before the Earl of Suffolk for a special address, all of them intrigued and some, a little nervous. As Henry studied the Earl's face, he could not help but notice tiny beads of sweat bubbling to the surface and how his jowls trembled as he imparted what he needed to impart:

"I bear news that is in equal measure unfortunate yet fortuitous dependent upon your viewpoint… I know some of you will be sad to hear that I am going to relinquish control of this most glorious House to a new owner."

The servants all gasped which Henry found strange, for they had been talking of nothing else these past few days. Most had guessed what the Earl had just said but no one had guessed what the Earl and his wobbling jowls were now about to impart:

"You should all be excited by the fact that the new owner is none other than His Majesty your King, which should go some way towards softening the blow..."

Servant Jack was first to show his cards: "Three cheers for the Earl – Hip, hip, hooray!"

The response from the staff was lukewarm until Henry declared his hand:

"God save the King!" whereupon the normally shy maids, porters and footmen exploded into dance. Henry was well aware of the Earl's eyes boring into his own but, as his future was going to be determined beyond the confines of this place, Henry felt he could witness the unravelling events as a detached observer. He had no doubt the opportunity would come again to impress his King – but he just prayed it would not take too long...

Despite having no belief in the Almighty, Henry's prayers were soon answered. On the third day at the third stroke of the clock a cry rang out from the third floor of the Manor releasing an avalanche of babbling servants, cascading down the stairs and rushing like a torrent along the corridors, knocking over furniture they had spent hours polishing only days before, to reach the windows overlooking the drive and obtain a glimpse of their new King. Engulfed by the wave of hysteria young Henry raced off to check all the windows, doors and locks for any squeaks or rattles he was responsible for fixing – so keen was he to impress the new King.

For a few expectant seconds, as the servants stared with nostrils squashed against the windows, nothing stirred, then finally a horse appeared from between the distant bushes, then another, both bearing soldiers resplendent in the style of the Royalists – not for them the stiff metal carapaces of Cromwell's forces but the resplendent outfits of the King's Men - inspired by the French Court of His Royal cousin Louis.

A cry suddenly rose from the crowd, "A coach! Here comes the King!"

CHAPTER 5
LONG LIVE THE KING

Captain Smith was a handsome muscular man of 29 – all you would expect of a decorated officer in the English Army, broad of shoulder and square of jaw. His voice conveyed authority with a smattering of good humour – the latter quality having connected well with the Merry Monarch:

"A fine choice of residence, if you don't mind me saying so Your Majesty - close enough to your Queen and Parliament, yet far enough away to profit from the abundance of country air after the ravages of that damned Plague."

King Charles was obviously relaxed in this Officer's company as evidenced by his intimacy in matters of the heart and mind:

"My dear Captain, the distance from those wretched Parliamentarians could never be great enough, but this house holds other attractions for me – Newmarket Racecourse lies but 25 miles to the North"

"You place bets on the horses, Sire?"

"Absolutely Captain – hereafter I decree it shall be named 'The Sport of Kings'. It is a foible that lightens my mood and unfortunately also my purse."

A dainty ditty suddenly interrupted the King, playing from beneath His jacket. Retrieving a small gold locket, the King opened it to reveal a miniature portrait of the actress Nell Gwynn,

"Moreover, my darling Nell lives but a short ride away."

Captain Smith smiled as he watched his King plant a kiss on Nell's lips, in the knowledge that were he ever to entertain thoughts of a coup, he had now heard enough evidence to bring about the King's downfall.

Two soldiers on horseback acted as front guard ahead of a dozen more seriously armed with musket, followed by 10 ponies carrying boys playing pipe and drum directly in front of the Royal Coach with a dozen more soldiers bringing up the rear - a Royal Train of sorts.

Henry watched in fascination as the House came alive like a huge machine of moving parts where every person knew their place - the staff lining up to greet their monarch all of a fidget and fuss - the cook, a proud king of his own domain, maintaining the appearance of calm despite having been caught off guard by the number of guests, ordering staff to gather rods and sprint to fish the river, while commanding others to raid all the known poachers' dens and steal back what had been stolen from the Manor in the first place. All the while, as his staff raced this way and that, the cook took advantage of his position, taking succour from the Sherry he had kept hidden in the cupboards for an occasion such as this.

The main doors opened wide to greet the King – Henry watching in awe as His Majesty flowed in like a swan leading his aristocratic cygnets whilst trumpeting Royal commands: "Replace those dull curtains at once. Have something gay for Heaven's sake – I shall need my guests in good spirits - the better to win them over! I compel you to be merry – fresh uniforms for all. Think of colour, gentlemen – look to the French style – Cousin Louis maintains a colourful and jolly Court. Opposition fails when jollity prevails!!!"

One elderly servant trying to keep up with the striding King, repeated the Royal Commands committing them to paper,

"'Opposition fails when jollity prevails'…Oh my goodness, whatever next…"

Of greater significance was the pretty young man striding enthusiastically behind the King whilst being introduced in the same vein,

"Welcome to my first Royal guest! Young Prince William has received an Honorary Degree here at Cambridge, making his uncle very proud. Prepare our Prince a meal fit for a King."

Applause broke out fuelled by a release of verve and enthusiasm held captive far too long under the rule of the King's puritanical father, and the ensuing stiff parliamentarian Cromwell who followed.

Fuelled by the same enthusiasm Henry rushed forward holding out the teapot and China cup for the Prince, but King Charles intervened,

"Beware of this young man William – he seems intent upon making an impression. I have carried mine here ever since we last met."

The King pointed to a blister on the end of his tongue in good humour and it was in that same humour that Prince William responded,

"So, I shall need to keep my eye on this one Uncle, for his eyes do twinkle so I confess I shall never forget them."

Henry stiffened, unsure how to respond until his King gave him a slap on the back and he received a wink from the young Prince, raising eyebrows from the other servants, in particular his young friend Jack…

Two hours later the jollity continued after dinner down, in the servant quarters with Henry, as usual, being content not to speak but to listen, and this time being the butt of all their jokes,

"Did you see that - Henry received a wink from the Prince…"

"And a slap from the King…"

"Aye, so Henry has left an impression on both Crowns…"

"Times are a changing. I hear *this* King is fond of merry making…"

"That makes a pleasant change…"

"I hear that is not all the new King is fond of…"

"Oh yes indeed, the beautiful Miss Gwynn!"

Henry's ears pricked up at the possibility of an inroad into the workings of the Royal mind and, if fate allowed, a chance to impress,

"Who is this Miss Gwynn of whom you speak?"

The servants joined together answering:

"You must surely have heard of Nell Gwynn!" said one.

Another added, "Her mother got so drunk she drowned in a pond…"

Henry laughed, completely captivated by such ungodly behaviour. His father would not approve but then, as his father was not there, who was to know or care? …

Jack leant in close to Henry's ear,

"She has played a fine game with our King…an actress who came from selling oranges on Drury Lane to becoming the apple of His Majesty's eye…"

The other servants joined in the jollity, "Even the great Samuel Pepys has fallen under her spell…"

"Many plays have parts written especially for her…"

"Aye, but her most famous role is that of the King's favourite mistress…"

"I have heard tell He has even had a tunnel built between Newmarket Palace and her house so he can visit discreetly…"

Henry was shocked at such behaviour, beyond anything his father would condone, leading him to be completely captivated and full of questions,

"But His Majesty is married – does not Queen Catherine mind? Surely it is against all the laws of God?"

The servants took turns to toy with him,

"I am sure she does mind, but The King is The King…"

"Sweet Nell has captured his heart though – I hear she means the world to him…"

Henry's mind was swirling, which did not go unnoticed by Betty the scullery maid,

"Come, come Henry, have you never experienced the sweet chase? The prize of a young girl's heart?"

Jack joined in with a mischievous grin, "And more if you're lucky…"

Henry was at a loss, feeling out of depth, wishing the talk was of Science or some other thing he could relate to,

"I can truly say the matter has never crossed my mind."

"That is only because it has never crossed your path…" stated Betty, unable to resist adding a sprinkle of mischief,

Jack interjected bringing a conclusion,

"Aye, experience another side of life Henry. Come into town this night and enjoy an ale with us…"

"My thanks, Jack, but I am really not sure – I have never had an ale in my life and moreover my father would not approve."

"Your father will never know," added Betty to the mix, "Since His Majesty came to the throne, we do this every week – it is the 'new' way…"

Sensing victory Jack moved in for the kill, "I have it on good authority that 'The Golden Goose Inne' has prospered from His presence on more than one occasion of late…"

Henry's mind was made up, "Well, if it is good enough for His Majesty…"

CHAPTER 6
EXPLORING A NEW WORLD

That evening, with his face hidden under a blanket, Henry climbed into the manor's cart full of servants, unrecognised by his father staring down disapprovingly from his bedroom window. Henry's excitement was tinged with fear for this was the first time he had ventured out on such a journey – an evening spent solely with young work colleagues to experience a slice of life he had never tasted before – an evening like no other and, being free of expectations, would also surely be free of disappointments...

Peering out from underneath the blanket Henry was agog at the sights, sounds and smells of such a busy market town, now bathing him in its glory - the hustle bustle of townsfolk relishing their weekly freedom from the shackles of work and woe for one special evening – Friday night in Saffron Walden.

Their young horse Toby came to an abrupt halt outside a public drinking house on the main street, snuffling its disdain at being made to stop so soon from enjoying an evening away from the confines of the Manor's stable. Understanding Toby's predicament Henry jumped down and put his arm around its mane, looked deep into its eye, and offered a gift,

"Take this my young friend. I feel you and I are riding the same journey - both being allowed out for the first time. If I am right, I shall be enjoying the taste of something special this night, and you deserve to do the same. So here, take this…"

Henry produced an apple and swore to his dying day that when Toby saw the fruit, he understood what Henry had said, before demolishing it between his massive teeth.

To Henry's amazement 'The Golden Goose Inne' was full to the brim with the faces of locals exercising their mouths by glugging ale, spouting songs, laughing and belching – in complete contrast to his father's pious meetings where mouths moved so little, whispering psalms and prayer, that they might as well be not moving at all...

So far, one face amongst the many had deluded him – the main reason he had agreed to join this merry group - the bubbling black hair, moustache and eyebrows of the man he wished to meet again – King Charles.

Jack approached carrying a distraction,

"Try this, Henry, – your first taste of ale!"

"My thanks, Jack. I would toast His Majesty but alas, I cannot find him."

"He doesn't come here every Friday, you know. Forget him this night – there are plenty of other good folk here deserving of your attention."

Jack drew Henry's attention to a family sitting in the opposite corner, a middle-aged pleasant looking couple, and more importantly, their young twin daughters.

As Henry sipped his first ever ale, he was surprised to note that those same twins looked more agreeable with every mouthful. He also noticed that with every one of Jack's mouthfuls, the sillier he became, laughing at the girls, pointing at the girls and going as far as to wink at the girls. Henry laughed more out of embarrassment than anything else, expecting the twins to take offence and leave the establishment, yet to his great surprise, the twins returned Jack's laughter, seemingly encouraging him. Henry observed this slice of human behaviour as he would a scientific experiment, intrigued yet detached, chronicling the events in his mind. After only five minutes Henry had surmised that the escalation of activity was being brought about by a catalyst - the action of a third party: a man with a hurdy-gurdy had fired up a popular tune of the time immediately raising the energy of the place, enticing increased laughter, much merry making and some dancing.

"What do you think of the music, Henry?"

"I find it quite fascinating, Jack."

"Fascinating? Does it not entice you to move?"

"I take it, you mean I should indulge in rhythmic physical activity? Not for me I'm afraid, Jack."

"Ha, stop being boring, come and enjoy…"

Henry laughed as he watched Jack, under the ale's influence, wobble over to the twins beckoning with arms outstretched and grinning like a silly little boy. Henry felt assured this was going to end in failure, but to his great surprise the girls reciprocated, rising to join hands and be led to a space on the floor where they all hopped, twisted and span, as if convulsing in some sort of well-rehearsed fit to the harmony of the hurdy-gurdy. As his only previous knowledge of communal music was standing perfectly still singing The Lord's Prayer, Henry found this whole experience quite bizarre – even disturbing, as sensible-looking folk got to their feet and began gyrating in time with the musical beats as if experiencing some form of fit.

Relaxing under the influence of more ales than he would like to admit, Henry eventually succumbed to the encouragement of the servants and, with all inhibitions lost, attempted a jig, catching the attention of a group of young women with whom he attempted to dance.

Making his way to the bar with a bellyful of ale and bonhomie, Henry made an understandable mistake for such an inexperienced drinker:

"Good landlord, permit me to consult my purse and purchase drinks for all my friends here be they servants of the Manor or these new-found beauties at my side."

But as they all ordered different concoctions Henry was thrown completely out of his depth:

"Good landlord, please could you supply me with some of your ales, five of them being of the Light variety and three of the Mild for these ladies here - whatever that is – I mean whatever a Mild is because I know what a lady is and none here are examples of that…"

The landlord was beginning to warm to Henry now that his money was being placed on the bar.

"…and a Stout if you will - although I have no idea what a Stout is, it describes the lady who ordered it very well…"

The landlord followed the line of Henry's finger aimed at a portly woman amongst the crowd and grinned – a secret joke between men.

"…also, a wine of your Red variety as well as one of your White. Quick as you can, good sir, while I am still in a fit state to pay for them!"

The landlord now proved he had a voice commanding enough to call out last orders above all the merry making:

"Be patient, lad, you'll get them as quick as I can pour them but as I only have two hands you shall have to be content to wait!"

Enjoying the banter Henry swiftly produced a reply:

"Then one day I shall conjure a machine that can supply all manners of drink that your guests require without need of assistance."

"Well, until that day comes, I shall just have to manage with the two arms God gave me."

The merriment continued and Henry grew increasingly intoxicated, dancing and singing with a proportionate decrease in ability.

Henry was too drunk to hear the twins, but if sober enough he would have heard one of them say…

'He is such a charmer…'

…and the other reply…

'And so handsome!'

Their banter continued beyond the confines of that inn, and beyond the limits of that day into the following morning and the confines of another fine establishment – 'Taylor's the Tailor Shoppe', wherein Jane was helping her mother whilst keeping one ear on the twins' conversation as they chose scarves nearby.

"He talked of things so fantastical – his head was so full of ideas I could barely breathe!"

"Yes, and such a catch! What with him and the King and everything…"

Mere mention of the King was enough to catch Mrs Taylor's attention,

"Who is this young man of whom you speak?"

The twins were only too eager to continue their praise,

"Henry – he works at the Manor…"

"And he joked all night about how well he knew the King…"

"And about how he is chasing fame and fortune!"

Mrs Taylor decided she'd heard enough,

"Really! And does this Henry possess a wife?"

"Not at all! And we shall see him again at the Golden Goose next week – he loves it so much he vows to return every Friday! I shall make it my weekly duty to attend!"

"And so shall I!" added her twin.

Every mother wants the best for her child and Mrs Taylor was no exception, willing to offer the most inappropriate advice to any woman trying to get in her daughter's way, and that day such a woman was standing next to her…

"I shall need to buy a dress for next Friday night – could you advise me on the most flattering style?"

The woman's twin was not willing to be left out, "Oh me too, if you please!"

Mrs Taylor glanced at her daughter with a conspiratorial grin,

"You can rely on me, ladies, let's see how I can best serve the situation."

And Jane observed, with some degree of admiration, how her mother guided the women to the most inappropriate dresses yet convincing them otherwise, fielding comments such as: "Do you really think I suit mauve?"

"Oh, most definitely, Madam…"

"My mother says mauve makes me look ill…"

"Quite the contrary, Madam, it makes you look ripe for the plucking…"

Mrs Taylor proved herself to be a perfect mother that day – her whole focus being on the betterment of her child for the benefit of them both.

If only more parents were of the same opinion, Henry might have been allowed to enjoy his young life a little more…

CHAPTER 7
A GOLDEN OPPORTUNITY

Every so often Fate provides a golden opportunity so fortuitous, it must be seized upon immediately.

At Audley End Manor the following morning one such opportunity was about to present itself to our young hero and, ever the opportunist, Henry was prepared to make the most of it...

King Charles was in his bedchamber choosing a wig - his face being powdered by Rupert, his most trusted manservant, – a lowly paid, but secure and prestigious position as long as one remained trustworthy, loyal and willing to supply His Majesty's every whim - another one of which was about to be announced...

"This periwig is My Lady Castlemaine's favourite. I trust you shall arrange it upon this Royal Head to best effect - thereby ensuring your Majesty shall plunder the treasures he desires this night!"

As the King's personal manservant Rupert was acquainted with most of His Majesty's private foibles and, moreover, knew how to keep them private:

"Your favourite Rodentia tonight, Your Majesty? The black or the brown?"

The new King was willing to be swayed on most things if persuaded otherwise, but on the matter of his hairpieces, King Charles proved to be unshaken,

"Most definitely the black! These scurrilous creatures shall witness whatever marvels befall my eyes this night!"

After having the rat skins placed upon both eyebrows to best effect, the King rose too quickly from his seat and, in so doing, accidentally, knocked his golden locket to the floor before stepping back with a sickening crunch.

Crying out in despair, the King attempted to open the dented cover and, in so doing, heard it emit the first three notes of the charming little tune trapped therein, that kept repeating themselves incessantly. The King was beside himself with grief, bringing the locket close enough to his face to notice the portrait of Nell Gwynn had been split with a crack across the middle of her face resembling a grotesque scar.

The King was inconsolable, "Oh, my dear Nell what have I done?"

In seeking solace, His Majesty decided that his sorrow would be best tended elsewhere, so whilst grieving the misfortune caused to his favourite mistress, the King sought solace in the arms of another.

Inquisitive as always, Henry had kept his eyes on the King's movements and after listening at Lady Castlemaine's door long enough to know the King would be staying awhile, tiptoed across to His Majesty's bedchamber and peered through the half-open door (for Henry believed that doors left half-opened always required one's full attention).

Rupert was studying the broken locket, shaking it, bending it, attempting various ways to mend it and win favour with the King, but soon realised it was beyond his meagre skills and replaced it on the table before leaving the room. Checking the coast was clear, Henry grasped this opportunity, crossed over to the locket, placed it in his purse, and swiftly departed.

Later that night, alone in his workshop, Henry cautiously attempted to open the broken locket. The tiny tune came as a surprise striking its repetitive soliloquy as, gently prising open the lid, Henry discovered the true extent of the damage to Nell Gwynn's image - a grotesque scar across her mouth.

Henry's mouth, however, formed a wide smile as he realised both Fate and Fortune could play their hand in favour of what now lay broken, in his.

Over the following three days and nights Henry summoned all of his skills to repair the King's locket, every night the state of repair being revealed by the extra notes playing until, on the third night, the tune was fully restored, and Miss Gwynn's mouth regained its former smile.

Henry smiled too, from the satisfaction of knowing that this small deed would bring him huge benefits, for he had discovered a powerful truth - the way to attract this King's head was to aim for His heart.

The dark foreboding clouds that bubbled over the buttresses of Audley End the next morning perfectly complimented the mood of the King as he climbed aboard the Royal Coach accompanied by his trustworthy bodyguard – Captain Smith.

"A week in Parliament! You know something, Captain – I would wager all of my subjects believe their King to be the most powerful man in the land – well, where is all that power now? I have no wish to spend one more hour amongst those insufferable politicians, but I am The King, and that is my burden. I tell you Captain; they are the ones rolling the ball and I am but the skittle."

Captain Smith was a man confident in waging wars on the open battlefield against an enemy he could see, yet, would be at a loss confronting the duplicitous nature of wars waged solely with words against an enemy hiding within the shadows of Parliament. He commiserated with his King but could offer only simple truths based upon his own straightforward experiences:

"Your Majesty, battle has taught me one thing above all else - your enemy should never be underestimated."

"Rightly so, Captain - the wood that makes the skittle can also make a spear."

As Captain Smith pulled open the carriage steps, the King was suddenly distracted by a sweet familiar tune emanating from an upstairs window – the distinctive melody of the locket now fully restored, "It cannot be!"

The King was drawn back into the house as if pulled by an invisible thread, passing Captain Smith who, ill-equipped to help in such matters could only follow while protecting the rear guard.

His Majesty glided through the House led by the delicate chords of the locket as if in a trance, discarding his hat scarf and gloves to be deftly cleared away by swift servants, who seemingly appeared from nowhere and to where they returned as if by magic.

Ascending the Great Stairway to his bedchamber, The King was like a moth drawn to light, following the delicate sound of the locket, not daring to believe what he knew must be true.

With eyes open wide His Majesty peered through the part-open door of his bedchamber, and gasped upon spotting the locket chiming its merry tune on his pillow. The King was reduced to a child bounding onto the bed, clutching his locket like a young boy reunited with a favourite toy, holding it close, regarding the face of his true love in all her perfectly restored beauty…

"SO!"

The word was spat with a degree of venom reserved for snakes and villains of the Shakespearean stage.

Taken by surprise the young boy became The King once again, and as such turned to face the sight of His mistress blocking the doorway wearing a countenance as stiff as her corset, "You have been back but 5 minutes and already your heart is with another – and on our bed to boot!"

As always, the King's response was delivered with tact,

"Pray, go no further – may I remind you who you target with your tongue, Madam. Presently I cannot muster the diplomacy required to deal with this situation – I shall have full need of that in Parliament."

"I have not forgotten to whom I speak – a man of sufficient wealth to pay for my compliance – through the card tables of Mayfair. Fear not, Your Majesty, I feel luck will be with me tonight – at least on the cards!"

"Aye – the last time you thought that, my purse grew lighter to the tune of fifteen thousand pounds…"

The petulant Lady pivoted on her outrageous heels, aimed her Italian Curved Points* at the stairs, and launched herself at the Royal carriage as a perplexed Captain Smith passed her rushing up the stairs drawing his sword to protect the King, "I am sorry, Your Majesty, I heard shouting…"

"Put away your sword, good Captain, the deadliest weapon that a woman carries is her tongue and besides, she is impervious to pain. Now talking of tunes…"

The locket continued its dainty melody as The King crossed to the window and witnessed, with some degree of satisfaction, Lady Castlemaine ejecting herself into the downpour below negotiating the puddles in her elaborate boots, before launching herself into the carriage and setting off in a flurry of whip cracks, hooves and screeches as the unfortunate steeds pulled the petulant lady to her favourite place of solace – Mayfair.

"Can I fetch you anything, Your Majesty?"

The voice of his trusted Captain had calmed the King,

"Actually, yes, you can – pray tell me the name of the man who gave me back my Nell."

And with that Charles tied the locket around his neck, ensuring his darling Nell rested close to his heart.

Captain Smith set off immediately to find young Henry Winstanley, but young Henry had long since disappeared on a mission of his own…

CHAPTER 8
THE CONQUEST

St Mary's Church clock tower was the tallest building in the small market town of Saffron Walden and one of the tallest points in a county made famous for being so flat, reminiscent of Holland a mere 200 miles away as the crow flies (presuming a crow would wish to fly there.)

On this particular day it was a seagull that flew over the clock tower on the morning after a night of consistent rain which had left rivers and ponds overflowing in its wake. The gull circled the tower a few times before deciding to land alongside the clock and rest awhile, and as it rested, far below in the street, a mother was racing on a mission to ensure her daughter would become a woman of substance...

The clock struck two, startling both the gull and Mrs Taylor – the gull taking flight as did Mrs Taylor scurrying to the door that could open such wonderful opportunities for her daughter Jane.

Her husband jumped as the door banged open, missing a stitch on the garment he was mending, "Is everything all right, my dear?"

His wife was pragmatic and in no mood to talk, "Where is Jane?"

"Following your orders, taking that dress to Miss Norton for pinning - she will be back soon."

At the sound of that name the blood left her cheeks, "Miss Norton – is she one of those who came in fussing about the young man from the Manor?"

"Aye, I think so my dear – whatever is the matter?"

Mrs Taylor couldn't have turned more ashen than if she was prone in her coffin,

"Oh, my goodness! Jane must get back quickly - that young man is on his way over here as I speak!"

Mr Taylor had never seen his wife in such a state, being a woman of business and a wife of control Mrs Taylor always wore the trousers in their relationship, always had a calm head and always a cool manner. His words emerged as a mix of trepidation sprinkled with frustration, "Is it important for her to be here? I can fix whatever he needs…"

"Believe me, you cannot. He needs what you cannot possibly give!"

"Oh – so what does he need if you know so much?"

"He doesn't know yet, but he will, soon enough! Oh, my goodness, I think that is him now…" Henry was tying his horse a mere ten yards away, "…please hurry, Jane or all will be lost!"

"All what my dear?" added her husband, vaguely.

"Oh, for Heaven's sake, be quiet you fool! Whatever did I see in you!"

"Her!!" he exclaimed, sitting bolt upright pointing out of the window, where Jane could be seen making her way towards the shop, side-stepping the puddles along her way.

Mrs Taylor raced over to join her husband, grabbing his hand tight upon spotting their daughter closing in behind Henry, then tighter still making her husband yelp as they both neared the door, wife and husband's heads locked together to see how their daughter would fare…

As Henry approached the door, Jane appeared in front blocking him, her face hidden by a hooded overcoat, stopping dead to avoid a sizeable puddle. Henry, in his haste, almost bumped into her,

"Please accept my apologies, Madam. My clumsiness is only exceeded by my inability to excuse myself sufficiently."

Brimming full of bravado, as Henry performed an exaggeratedly low bow, his gaze fixed upon the clear reflection presented in the puddle - this lady's naked legs. Making the most of the situation, Jane pulled up her skirt to reveal a little more.

Rooted to the spot with heart thumping against his chest as never before, Henry's eyes voyaged upwards taking in the beauty of Jane's form until, arriving at the hood, he discovered the treasure he sought, for when it was pulled back like a curtain on the stage, it revealed the most sensual face Henry had ever seen. As is so often in circumstances like this, Henry's cheeks suddenly reddened and his neck pulled tight,

"Could I at least offer to escort you safely to your destination, Madam?"

This face of beauty created an equally beautiful smile,

"You may, sir."

Young Henry was completely transfixed, stiff as a rod unable to perform the most basic functions of polite society, so if anything was going to develop from this situation, it would need to be instigated by the young lady. As has been the case so many times before, the lady took the lead by offering her hand, and the gentleman obliged by offering his arm, so history could repeat itself at 14:15 on the afternoon of May 1st 1661 by the joining of these two young souls in the dance of life.

"Your humble servant, Madam."

No sooner had Henry stepped towards the shop door, than he felt the shock of her tiny cold hand on his, and she took the lead once again pushing him back sufficiently to spin on one foot and hop into the shop with the other.

Henry, bemused at the situation, followed her inside with much less grace than her, to be greeted by an ingratiating bow from Mrs Taylor.

Delighted at being treated with such dignity, Henry responded with an even more ridiculous bow, "Madam, this delightful young lady has guided me to my destination when my intention was to guide the young lady to hers. I am intending to purchase garments suitable for a meeting with The King, and would be grateful for your advice on the matter."

Winstanley's Wonders

As if mention of The King had set her in motion, Mrs Taylor suddenly sprang forward with an agility defying her years landing precisely on the spot where she commanded the paths in all directions, and made the most of her achievement by spreading both arms to maintain that control.

"We are honoured by your visit, sir. The young lady to whom you refer is my daughter Jane - I can recommend her as the most suitable in the whole county, nay, even the whole country, to address your needs."

"Then I am indeed fortunate for having made your acquaintance Madam, and declare myself completely in your daughter's charge."

At this point, Henry was so completely transfixed by the young lady standing within touching distance that, when her eyes looked directly into his and her lips addressed him directly, he was like a fool barely able to string two words together. Her neck begged to be stroked by his hand, her hair for the touch of his fingers that when she broke into giggles, he released a laugh as silly as any child's.

"Forgive me, Madam, you have reduced me to a child."

"I beg of you, sir, please call me Jane."

"Then Jane, I implore you to call me Henry, named after my father, Mr Henry Winstanley."

The name drew Mr Taylor from his desk,

"Your family is well known to us, sir - I remember your father as Master of the Alms-houses before being made Steward of Audley End. We have supplied various garments for him over the years."

Henry smiled, "Then in some measure you must know him more intimately than I."

Never the shrinking violet, Mrs Taylor could not bear being excluded from the conversation,

"He cuts a fine figure…"

Her husband joined the banter, proving a fine match,

"Six foot and one-half inch…"

"…prefers a jacket loose across the chest…"

53

"Forty-four-inch mind you…"

"…trousers wide across the waist…"

"…thirty-two inside leg…"

Henry's laughter burst forth full and true,

"Then indeed, you both have the measure of my father!" and turning to Jane, "Could you please transform the humble figure standing before you now? I will place myself completely at your mercy!"

Moving things on, Jane produced a winning smile,

"Then please step over here, sir - I have some jackets in mind that will suit you well."

"Take heed, Madam – the next man to see what you recommend will be none other than the King of England himself! I hope to create a favourable impression."

Henry was thoroughly enjoying their good-natured banter, in equal measure to Jane.

"Place yourself in my hands, Henry, and I promise you shall be left completely satisfied."

For the next two hours Henry did just that – placing himself entirely in Jane's hands. The first hour was filled with her measuring him, a thrilling experience for a young man who had never felt a woman's touch before, trying on garments of vivid colours and styles, he would never have had the inclination to wear before and being mocked by his own reflection for doing so.

His eyes barely left Jane throughout, of which she was obviously aware,

"You are making me nervous, sir. Pray, what are you doing?"

Henry replied with confidence, "I am looking at you."

Such a direct response delivered with such conviction caused Jane to blush - her cheeks swelling rouge red, and as she smiled, her cheeks seemed to trap the blood all the more,

"You should be looking at yourself, sir."

"But why, when I can look at myself any time I wish?"

"You cut a fine figure, sir."

"I would say more like a fine peacock, madam."

"Do you not like what I am doing for you?"

"I have never seen myself looking so elaborate or feeling so elated for being so. I declare this to be my choice of apparel from this day forth, and for you to be my personal dresser."

Henry stared entranced at this delectable creature as she struggled to regain her composure, "I beg you, sir, do not mock me…"

"I could never mock such a captivating product of nature, Madam; indeed, I must confess to feeling strangely hot myself."

"Then should I open the door for you?"

"Absolutely not, I am enjoying being caught in your spell."

"I am glad of it."

"Then consider this proposition – as you have succeeded in both making me look and feel like a new man, can this new version of me have the honour of seeing you again, for I suppose there are many more alterations that could be made?"

Jane's reply came swift and true, "Of course, you can, I would love to."

"In that case would you consent to a ride? – My best ideas come when I ride."

"Then I should very much like to share that experience with you."

Henry paused for a moment, full of self-doubt, wondering whether to dare reveal his greatest fear,

"I feel it is only proper to reveal that some people consider my ideas whimsically immature. I shall not take offence if you think likewise."

"I find most men to be immature, so worry not."

Henry competed with her blush admirably,

"I have to confess the prospect of meeting my King now, excites me somewhat less than the prospect of our ride!"

As Henry walked out of the shop Jane's voice made him pause,

"Henry, please remember to tell me of your news with the King."

"I shall do, Jane – on our ride!"

Both had exchanged their first names again – inconsequential in an everyday conversation, but to Henry this familiarity felt intensely intimate and therefore, so very exciting.

With renewed gusto Henry aimed his horse for Audley End, his mind completely occupied by this captivating woman.

CHAPTER 9
A CLERK AMONGST CLERKS

Being ever the optimist, Henry felt that at last his journey to Respectability had begun. A meeting with The King of England secured his place on a par with dignitaries, nobility and, most importantly for Henry - the respect of his own father...

The following morning all the mirrors between his room and the Great Hall were filled with Henry's reflection, checking his appearance: Henry outside his room, Henry two minutes later at the top of the stairs, Henry three minutes after that at the end of the Great Hall and six minutes after that outside the doors of the Great Room, inside which the King of England, Scotland and Ireland awaited him.

Henry wanted to not only look *his* best, but the best of all the men there - excepting the King of course...

Putting all fears behind him, Henry should have been relieved upon opening the huge doors to find The King absent, and a sombrely dressed man in his place, but disappointment overwhelmed him.

In place of the King, Henry stood dressed in all his finery, facing a middle-aged officious-looking fellow staring straight back, whose hair, complexion and eyes matched the grey of his attire. When the man spoke, his words were coloured with a soft Edinburgh accent as lifeless as the rest of him:

"Mr Winstanley, permit me to introduce myself - my name is Mr Gray. I am the Head Clerk acting directly for His Majesty's pleasure, and it is in this capacity that I bear good news for yourself and your family. In His generosity our gracious King has recommended your position to be raised to Clerk of Works. Why His Majesty has chosen to elevate a junior to my status is known only to Himself - a Clerk in His Majesty's service -

an achievement for which I have aimed all my adult life and of which I am justifiably proud. (At this point, he made a point of sniffing into a 'kerchief which Henry took to be a pause for dramatic effect.) He commanded me to present you with this pocket watch. He is fond of timepieces and informed me you would find it somehow appropriate as His likeness is displayed upon the back."

With some ceremony the clerk presented a gold case containing a pocket watch attached to a necklace. "I believe His Majesty shares your interest in timepieces."

And with a shake of the hand and a smile as smug as his bow, the Clerk took his leave.

No sooner had that self-righteous presence left the room than another appeared, in the shape of his father.

"You have been presented with a great opportunity. How you deal with it shall determine the rest of your life. Do not waste it…" adding as he was leaving the room, "You look over-dressed and somewhat foolish in that respect. Remember that, to be taken seriously one must practice restraint. Good day to you Henry…"

Faint praise indeed but knowing there would be nothing else on offer, Henry resigned himself to face disappointment alone in that huge room watching the trees on the drive outside shedding the leaves they no longer cared for and, understanding their predicament, feeling sympathy for them.

However, the optimism that had helped him survive the recent past returned to offer support again. Remembering the King's gift, Henry opened the gold locket to be faced with a likeness of His Majesty supported by a gay tune intended to lift the spirits. It worked for a few seconds until, upon turning to leave, Henry caught sight of his own reflection and for once, had to agree with his father - he indeed looked overdressed.

He thought about it for a minute, and after 61 seconds decided not to care what others thought and bounded away.

Thus, by the shaking of hands the dour old Clerk was, in essence, congratulating himself, leaving the room for the new dandy Clerk to take over and begin inching his way closer to those of wealth and power...

CHAPTER 10
SEIZING OPPORTUNITIES

The House of Commons was full to the brim with men making a lot of noise about things they had far less knowledge about than they ought. Amidst all the mayhem the one man who should have been carrying sway over it all was swaying with frustration, wanting to increase investment in an area others found inconsequential. Extracting a pocket watch from the most apt of places – his pocket – King Charles attempted to gain support for his cause.

"Gentlemen, Great Britain was aptly named by my father, for Britain is made great by its people, and the British people's greatest attribute is innovation – from the skill of our boat builders in creating the greatest naval force in the world, to the training of our armies to maintain it, thus spreading Great Britain's influence across the world and benefiting from the subsequent trade - tea from the East, cotton from America and silk from the Orient. The word 'Great' also applies to all of you seated here under this greatest of roofs - the world's first truly democratic Parliament ensuring that those who govern the people are enrolled by the people for the benefit of the people. We all remain grateful to you, gentlemen."

A smattering of applause spread across the House from manicured hands that had never seen any form of manual trade save for the signing of forms. With the House in such a self-congratulatory mood the King seized the opportunity to promote his own personal interest,

"Science is where we need to spend our purse, gentlemen, or risk falling behind others. It took an Italian to explore the vast ocean of the heavens just as the Vikings sailed beyond their horizon to plunder gold, we British ventured to far off lands to bring gifts – the tobacco we now smoke, the potato we now eat and more recently the tea we now drink from China cups. I firmly believe our horizon stretches to the heavens and it is our duty as Englishmen to exploit it, then protect all that we have exploited. I propose a telescope should be built at a place close by, enabling men of Science to explore heaven's riches. As Patron of the Sciences, it is in the national interest for me to have a telescope installed at Whitehall, so that I can keep abreast of their discoveries."

The King's speech was falling on deaf ears, for political heads were not notorious for their imagination. Surveying the surrounding disinterest, the King knew he had to attract their interest somehow and thought patriotism worth a try:

"Innovation is this country's greatest strength, gentlemen, from Charles Mason's pocket-watch nestling in my jacket to the Naval Warships of John Hawkins and the three-deckers of Peter Pett – forming an unbeatable naval force – the envy of our enemies and the means to finally defeat those dastardly Dutch."

A plump MP who looked somewhat Dutch jumped to his feet, jowls quivering, to address the House in a clipped accent betraying his Germanic roots,

"His Majesty's hunger for War with the Dutch is not in England's best interest and moreover, He does not have the funds to support it."

'Not having the funds' was a phrase the King was growing weary of for not only was it true, but it showed he was at the mercy of those he despised.

'Those he despised' fell silent as soon as He rose to speak,

"I have news of which England can be proud - this day The Senior Service, my most Royal of Navies, has captured New Amsterdam from the Dutch on the Eastern Coast of the Northern Americas. I decree it shall be renamed after my brother James, Duke of York. I consider 'New' York shall prove to be a great prize for this land."

The cheers he had hoped for never materialised – the elderly ministers electing to grump instead like a herd of swine uttering such negatives as; 'Never '- 'Madness' - 'Ridiculous', and the like.

"If I am proved wrong, I shall eat my hat, but if I am proved right the Dutch should eat their Prime Minister."

The House broke into laughter prompting the King to scan the members, observing who was joining the laughter, and more importantly, who was not - those faces He noted for future reference.

The unspoken rules of the House deemed that once a Minister rose to speak all others should refrain from interrupting, born from a sense of decency and respect. However, since few of them were decent and even less deserved respect, when another MP rose to speak, his words were interrupted by squabbling from the back benches,

"Members of the House, I appeal to your common sense and to your purse - we cannot afford any more conflicts with the Dutch; they would sink our economy along with our navy. The cost to the State's purse would be extortionate. Therefore, I propose a Treaty to divide the shares of world trade equally between both parties. The alternative would undoubtedly lead to higher taxation."

His message struck a chord with the Ministers, especially the threat to their purse, for self-preservation ranked highly on their list of importance. Drowning in a sea of 'ayes' the merry King sat fuming sensing his opportunity for a personal telescope dwindling. Amongst the mayhem, few noticed the young messenger who arrived unannounced producing a small piece of paper that was passed from his slender hand through many plumper ones until arriving at the King's velvet glove where it lay untouched for 37 minutes until His glove broke its seal, His eyes scrutinised its contents and His lips formed a smile.

Hiding His satisfaction, the King rose to his feet with a cough that silenced the MPs in an instant like a mass of squabbling schoolchildren once the teacher arrives in class. The patriots amongst them hoped The King was about to bear good news for the Nation, but many who had pockets grown fat on the bribes from her enemies hoped for favourable news for their paymasters instead.

Pausing for effect, The King strode slowly down the centre of the House lifting the tricornes off MPs on both sides whilst retaining his own.

"Gentlemen, I bring grave news…"

As he spoke The King inverted each hat so that anyone with half an imagination could tell that they represented ships to illustrate His news (yet because most had no imagination at all they were oblivious of everything).

"Whilst you deliberate what to do about our neighbours, they have come knocking at our door. The Dutch have sailed up our Greatest River to deliver the Greatest defeat to the World's Greatest Navy - and not in some far-off land but thirty miles from this House! Thirteen of our ships have been lost at Chatham Docks."

The King slapped the bench hard shaking most of the MPs and waking the rest to witness Him swiping thirteen hats to the floor before crushing them one by one.

"They arrived unopposed as if by a foregone conclusion engineered by those who wish our country harm. Look to the man sitting next to you, gentlemen, for there are some amongst you who are spies, and I shall deal with them accordingly. Treason is the most treacherous of crimes attacking the country that feeds them. Looking at each of your faces now, how can I be sure your surprise is genuine? I call upon all of you to look at your neighbour for any sign of treachery, for I promise those who act against this country shall lose all their worldly goods followed by their heads."

The reaction was instantaneous – all the MPs turning to face one another for any clues of dissent, thus the King had successfully spread suspicion throughout the House, creating the perfect scenario to rally them under one cause.

"What is more, those dastardly Dutch have plundered the very ships we plundered from them in the first instance! They are proving so successful I suggest that far from fighting the Dutch we should imitate them!"

Playing to a captivate audience the King turned his own hat upside down and pierced it with his feather like a sail.

Holding it out before Him the King made for the door,

"Their final insult is in taking 'The Royal Charles' as a trophy. I expect every patriot amongst you to unite in exposing our enemies or be considered an enemy yourself and face the loss of your head."

Upon reaching the door, the King beckoned Captain Smith to join him along with a dozen of his men who all drew swords and turned to face The House as a united front.

"So, I put it to you all, gentlemen – are we prepared to do nothing but politely plead for concessions with these scoundrels, or shall we dig deep to finance a war and bring the Dutch to their knees!"

The response was thunderous as every MP, aware of the guards' stare, rose to their feet in support of the King thereby drowning out his following words, "Your support ensures we shall have the ships to defeat the Dutch and with the grants you are about to sign, my telescope can be built ensuring England will not only rule the seas but the stars beyond."

Thus, the King left Parliament that day richer than when He had arrived only hours before: – England had seized New York, and its King had acquired a 36-foot telescope for His Privy Garden in Whitehall.

New York did prosper, The King did not eat his hat and eventually the Dutch did eat their Prime Minister...*

CHAPTER 11
OH, FATHER, LEAD HIM INTO TEMPTATION

*4**8 miles away a young man with desire in his heart and one woman on his mind, was about to be presented with the chance of another…*

Henry was in his workshop flitting from mirror to mirror trying on differing shirts, wigs and jackets including his father's, so obsessed was he with making a good impression on his first date,

"I must say you look ravishing in red, my love; do you think red might suit me too? Then we could step out as a matching pair, ha…"

Being lost in memories of Jane – her cheeks, her neck, her mouth, her laugh - Henry was completely unaware of another pair of eyes scrutinising him from behind the curtains,

"You suit red well, sir."

Henry leapt like a Spring Hare having been discovered by a stranger in his most private of places. Only five words had been spoken yet they defined the character of the woman delivering them, such was the dullness of her tone,

"Hello, sir."

She was slim to the extent of skeletal resembling a shirt hung over a chair, watching him with emotionless eyes that lacked any of Jane's sparkle, creating the impression of a schoolmistress regarding a mediocre student. She spoke again revealing no emotion whatsoever, with a dryness of throat that begged for a good cough,

"Your father told me to find you here."

"My father?"

"He is being very kind to me."

"That does not sound at all like my father."

It was only at that point that Henry noticed her neck for it was not a particularly attractive sight. Unlike Jane's voluptuous décolletage, the one facing him now was coated in alabaster skin, hidden behind plain white cotton as if curtains had been drawn across any semblance of womanhood. Something caught his eye though, glinting from behind the folds of that curtain in front of her thorax, of the same shape his father often held in his hand: the sight of the cross immediately put Henry on his guard, symbolising those soldiers of Cromwell encased in armour like metal beetles each of whom wore a cross - the symbol of a religion that made people feel guilty for enjoying the natural pleasures of being human.

Once again, the dry voice spoke,

"He is worried about his son as any good father should."

Henry replied, attempting to hide his irritation, "You are being presumptuous concerning the goodness of my father, so if you'll forgive me, I have an appointment to attend."

"With whom might I ask?"

Henry froze - the voice of a man had entered his room, an unwelcome voice which he knew well. A movement reflected in the mirror drew Henry's attention to his father stepping from the shadows to join Miss Worthy, giving Henry the impression of twins, such was the likeness of their stance and pious demeanour.

His father moved further forward to speak,

"Henry, welcome Miss Grace Worthy. She is proving to possess a purity of soul in keeping with her calling. She has already pledged to carry out the duties of this house."

"I do have to leave now, father."

"To visit Jane Taylor?"

Henry froze on the spot as if a stalactite of ice had dropped, pinning him to the ground. The following words arrived swiftly behind, like reinforcements in a nasty assault.

"Jane Taylor is a plain individual of limited calibre, possessing few skills of note and no leanings toward God, so therefore unworthy of sharing a future with my son."

Henry remained staring at two faces of fixed expression – his father's as if chiselled in stone alongside Grace Worthy's looking like a Mother Superior beyond any hope of a smile.

Henry's father took his leave delivering a nod that hinted his son should prosper from being alone with Miss Worthy.

Henry was trapped, desperate to escape all this worthiness and flee to where his heart was leading. Thoughts of Jane absorbed his mind – the way she moved, the way she felt, the way she smiled and the way she smelled, all conspiring for him to make his excuses and sprint for the stables.

Left alone in Henry's room, Miss Worthy took the time to observe the surroundings in an attempt to understand the workings of his young mind - the better to capture him. Yet failing to fathom any of his creations, she resolved to forget his mind altogether, and set her sights lower to arouse his body instead. Upon hearing the sound of approaching footsteps, Grace Worthy released all of her clothes save the crucifix around her neck, ready to commend her body into his hands. But when the door opened, the Winstanley who entered was not the Henry she had expected. At first Mr Winstanley senior took a step back but, upon gazing at the body God had gifted Miss Worthy, was reminded of the Book of Job, and in particular, the passage mentioning how God the Father could do whatever he pleased – so, being a father himself, Mr Winstanley took the Good Book's advice at its word…

CHAPTER 12
NATURE DECIDES

The very nature of what stirs a young, red-blooded male into action was stirring within young Henry that afternoon, powering his legs to sprint along the drive, race across the lawn and vault over the walls ready to leap onto a saddle and gallop away - but life being what it is, all the experienced horses were being employed elsewhere leaving only young Toby, restless and shackled to a cart as Henry's sole means of escape.

It was also quite natural that as a young woman, Jane would be harbouring the same desires.

Back in Saffron Walden, Jane's mother had her own battles to contend with as she fussed around her daughter making last - minute adjustments to her dress in order to lure the young man from the Manor, with his links to the King, and therefore the promise of what the future may bring for both of them...

It was the June solstice - the longest day of the year, yet despite the shop having no customers, it was as frantic as the busiest day of the week.

As usual, Jane's mother was in a practical frame of mind:

"What can that young man be thinking? Does he not realise the implications of closing shop on such a profitable day for just one customer? I have stocks full of the lightest of dresses on this longest of days. We shall lose money for every minute we remain closed."

Her father was equally nervous, "Absolutely, my dear, and there is no guarantee he will come anyway. He will have many young ladies following the trail of his scent, like bees to honey. We can only hope that Jane has left a lasting impression."

"She is my daughter - of course she has left a lasting impression, you silly old fool."

That 'silly old fool' kept a watchful eye on the street throughout, as he paced to and fro, supplying his wife with her needs – namely, buttons, needles and thread. But as time passed by, along with many potential customers along the High Street, even he began to lose hope.

Jane tried to retain her composure but could not help glancing up at the church clock every few minutes, dreading the movement of its hands, for every minute that passed made Henry's arrival less likely.

Her mother was struggling to remain optimistic for she knew an opportunity like this comes but once in a lifetime, "If you are to ride then you shall need our most fashionable boots – just try to keep them clean so I can place them back on display the morrow."

"Do not worry, mother, I shall look after them as if they were my own," came Jane's practical reply proving this apple hadn't fallen far from its tree.

The clock struck three, shocking the seagull snoozing alongside, sending it screeching from its nest. Mrs Taylor sighed, Jane sighed, yet Mr Taylor laughed, pointing at the window,

"You will not be needing riding boots today, my darling!"

Jane span around ripping the needle from her mother's grasp, spotting why her father was so excited - Henry was approaching in the Manor's cart. Jane was overjoyed, but her mother couldn't hide her disappointment,

"Goodness me, I would have hoped for a coach at the very least."

Ignoring her mother Jane laughed, crossed to the kitchen, grabbed the bottle and glasses she'd prepared for the occasion and rushed out to meet the delighted Henry, leaping into the cart revealing mismatched boots to her mother's disdain,

"Goodness! Look at her boots – what on earth will he think of her? What on earth will he think of us?"

As the happy couple headed off down the Main Street, the final words were left to the 'silly old fool' inside,

"Did you not see how he looked at her, my dear? Trust me – he will not be looking at her boots."

With Henry at the reins, 'happiness' became the order of the day - Toby happily pulling the happy couple along the track leading towards the River Cam, dispersing clouds of colourful insects attracted by the most glorious of suns.

And in the happiest of moods Henry waxed lyrical,

"Such a perfect day, the longest of days where the sun lights the world like an everlasting candle, enticing the insects to come out and play."

Jane sat transfixed by this most unusual of men, speaking of things beyond the subjects of farming and ploughs she was accustomed to from the local young suitors.

A large wooden fishing rod had been strapped vertically to the side of the cart, attracting much laughter from the young woman,

"My goodness, Mr Winstanley, what makes you think I should like to fish?"

"Because, Miss Taylor, I believe you are a girl who knows what she wants, and is prepared to use any bait to obtain it. Therefore, it stands to reason that Miss Jane Taylor and fishing are made for one anoth-"

A sudden jolt from the cart stopped Henry in his tracks, having unwittingly led the inexperienced young Toby into a dip, leaving him struggling to pull them out.

"Now I believe I know what kind of man you are, Mr Winstanley..."

Henry was struggling with the reins, "Indeed?"

"A man who needs steering on the right course."

To Henry's surprise, Jane took control of the reins, and succeeded in calming Toby sufficiently to hoist them out of the dip and pull them on through the woods.

Henry was delighted, "Thanks to you, the situation has improved to my advantage. I can now sit back and admire these wonderful birds."

Jane added playfully, "Indeed – then I am glad to have been of some assistance to you, sir."

"Have you noticed the birds in all their splendour, Miss Taylor?"

"You mock me, sir, I have seen birds on many occasions."

"But have you noticed how they refuse to fall to Earth, no matter how much the branches sway? Have you ever wondered why those branches sway so much in high winds but do not break?"

"I confess I have not, but I feel you are about to enlighten me."

"I have pondered on this many times, and concluded that the trees' flexibility is their strength. Bending to a superior force is wiser than standing against it."

Jane laughed at that notion, "In truth, my father's marriage survives by the same principle."

But Henry hadn't heard, his mind already drifting, "Pliability - I feel sure one day that conclusion will serve me well."

Jane steered the cart to a halt by the river snapping Henry out of his reverie,

"My thanks, you have found the perfect spot to conclude my hypothesis, for today this length of willow will prove my point."

As Henry grabbed the fishing rod, Jane stepped into action carrying her bottle and glasses over to a pleasant clearing on the bank and, presenting the bottle said, "I have contributed something for the occasion. Allow me to present Lemonade! – a gentle drink of water suffused with the juice of lemons - I hope you like it."

"Splendid – I have never before made its acquaintance. The fish can wait…"

Henry laid back, arms behind his head, soaking up the idyllic setting, as the rippling of water harmonised with the crackling of willow to create a serenade of Nature for intimate conversation - an opportunity not lost on either of them…

"From this day forth, I declare my favourite drink to be lemonade! I am indebted to you, Madam."

"I always bring some here."

"And do you come alone? Or is there always a lucky gentleman to accompany you?"

"No gentleman. I come alone and am quite content to do so - it gives me the opportunity to think."

"And may I ask where your thoughts take you, or is that prying too much?"

"You may ask, but you will find my response quite dull."

"Nonsense, Madam, how could I find anything about you less than completely captivating?"

His persistence was flattering and well appreciated, "Well, if you must know, my life skirts around numbers to such a degree, my mind has become their servant: the number of stitches on a hem, how many reels of cotton for a particular garment. I count how many we use each week, then try to think of ways to use less. My parents put their faith in the ways of old though, and are quite immovable on the subject. I so detest waste…"

Unused to sharing such personal thoughts, Jane was not surprised to notice Henry's attention drifting, as if in a dream, "…I am sorry, sir, I did not intend to bore you with my thoughts."

To her surprise, Henry sat straight upright targeting her eyes,

"On the contrary, Madam, I admire them. I must confess to my mind being like a bird – a young albatross in constant flight, searching for a rock upon which to rest."

To Jane's delight, Henry brought his face very close to hers, and tapped the top of his head, "I have such wonders stored in here, that the world will marvel at how they could all come from inside just one head. I do not fear death, only the chance my wonders will never see the light of day. Look here -"

Henry sprang to his feet, grabbed the rod, and drew an unusual pattern on the riverbank, "This is my design for a contraption that has been occupying my mind for the last few months – a device to turn water into wine, and more."

"The Bible called that a miracle."

"Modesty prevents me from agreeing – but I would settle for naming it a 'Wonder'."

"Henry Winstanley's Wonder – it has a good ring to it."

"I dare not tell my father of such things, for he would consider me blasphemous, and his wrath would be unbearable…"

It was now Henry's turn to notice someone else's attention drifting…

"A thousand apologies, you must think me either a fool or a conceited bore. The last ears to hear my dreams belonged to my father, but I know he considers them of very little worth. I can never please him. A simple pat on the back would suffice but he cannot even bring himself to do that – one father's hand upon one son's back…"

"I am sure your father is very proud of you. He should be. Your thoughts are unusual, unlike any man I have ever met. I do not profess to understand you, Henry, but I do believe in you."

At this point Henry dared to move closer, targeting her with the look from his eyes, and the words from his lips that came straight from his heart,

"Jane, you are the only person I have ever talked to about this – it has been held captive inside my head for many months now. It must have weighed heavy, for now I feel as light as the afore-mentioned albatross."

As Henry's face grew close enough to feel her breath, the sparkle left Jane's eyes, being replaced by something much deeper,

"I suspect any attempt to harness your mind would be met with such resentment, I could not bear the responsibility of causing it - so I say dream on!"

His face had arrived so close their lips were on the verge of touching but, having never kissed a woman before Henry chose to defy rejection, and swiftly withdrew choosing cowardice over valour,

"You say you have come here before, but never fished! That outrage shall be remedied forthwith. With this length of willow and this hair of horse, I shall attempt to be the huntsman gathering my Lady's food for this night."

As Jane sipped lemonade watching from the riverbank Henry carried his 12-foot rod across to the riverbank, where he began using it as intended, swooping its artificial fly across the river, to kiss its surface, enticing any hidden fish below to rise and take the bait.

Within seconds, and unbeknown to Jane, Henry spotted a sizeable trout swimming against the current allowing him to cast the fly ever closer whilst whispering a message:

"I beg of you, good Mr Trout, come over here and help me - take the bait but offer a slight struggle – not so much that I lose you, but enough to impress the young lady sitting over there. I'll make it up to you, I give you my word."

Engrossed in his endeavour to capture both the trout and the girl, Henry carelessly whipped the fly so far back it caught Jane's jacket, tugging at a piece of her fabric sending it flying back across the water. Henry was about to apologise when the trout suddenly took the new 'bait', leaping clean out of the water. Henry thanked the trout and beckoned Jane to help him land it.

"I am so sorry about your jacket, but I believe we have made the acquaintance of a trout with impeccable taste - he prefers your tweed to my miserable fly."

As she stepped towards Henry, Jane scooped a handful of water, and threw it at him laughing. Taking the bait Henry returned two handfuls, drenching her white cotton top. Apologising immediately, Henry was surprised when Jane didn't seem to mind at all, despite the water having clung to her top leaving little to his imagination. Thus, Henry was learning an early lesson about young maidens, and was about to learn a great deal more once Jane strode over to him, for her breasts were clearly visible, pushing hard against the drenched cotton top, and what excited Henry the most was knowing that Jane knew he knew.

Spotting his dilemma, Jane seized the initiative, pulling him close for a kiss. To her disappointment, Henry resisted her advances,

"Does it not concern you that the touching of lips has been outlawed - for the prevention of another Plague?"

But Jane's determination was unstoppable,

"So, tell me, Henry, which fool passed such an idiotic law?"

"Why, His Honourable Personage – Lord Falconer of course."

Tasting victory, Jane eased in closer for that first kiss, "And tell me to whom does that Honourable Personage answer?"

Stepping back Henry felt a tree trunk preventing his further retreat so tugged on his necklace, and held up the King's face as a shield - engraved upon the gold pocket watch,

"Well, the King, of course."

"And have you not heard of our King performing the kissing of lips?"

"Aye, I have been informed He has done so on many occasions."

Jane had arrived so close now that Henry could smell the sweetness of her breath,

"And being His loyal subjects, should we not follow His example?"

Having nowhere further to go, Henry graciously accepted defeat,

"You are absolutely correct, Madam."

"Then permit me to introduce something new into your life," adding, as she turned The King's face away, "Avert your eyes, Your Majesty."

Having had more experience in these matters than the virginal Henry, Jane loosened his trousers and took the lead in losing his adolescence to the first pleasures of manhood – a journey made but once in a lifetime…

…A while afterwards, feeling at one with the world, Henry fulfilled a promise flipping the trout back into the river, "My thanks, good Mr Trout."

The young lovers lay down together with Henry practising his new-found skills continually, throughout the rest of this longest of days. Both of them had been mesmerised by Nature – its forces that had joined them both together, as surely as a cock to a hen and the sight it had produced for them now – the final glows of a flaming Sun sinking below the horizon.

The tree they had made love beneath carried a simple message carved into the bark – 'HW' above 'JT' - a simple form of message used many times before to declare a love to the world that would last for ever – or at least for the life of the tree.

Back at Audley End that night, Henry slept better than he had ever slept before, fully quenched in body and spirit having shared both with this most wonderful of women, finally having found someone with whom he could trust to share his dreams – yet as he slept, Henry was unaware of another person in the adjoining room who was also interested in his dreams, searching his most private space where his construction of the planets lay hidden beneath a sheet.

A black jewelled glove snaked across the room homing in on the sheet before swaying to and fro like a snake considering its prey then lunging, grabbing hold of the sheet and tugging it off exposing the brass planets beneath in all their magnificence.

It proceeded to slide across the room pulling open drawers, reaching inside and finding sculpted pieces of metal waiting to be attached amongst other mechanical devices of mystery.

The glove raised a finger to its owner's lips stifling a laugh,

"What a useful young man..."

But a second intruder had just entered the room upon hearing the first,

"I am surprised you think so..."

Taken by surprise the glove knocked one of the metal pieces to the floor as the first intruder span around to face the second,

"Forgive my intrusion, sir, but curiosity compelled me to discover what was behind this closed door."

Henry's father gasped, "I heard noises from below but did not expect to find Your Majesty interested in my son's frivolities, my expectations of which always far exceed the results."

King Charles exchanged his smile for a frown,

"Sir, you underestimate both your son and your King."

Henry's father took a step back, unused to being chastised by any man let alone a King he considered to be a nincompoop*, "Your Majesty, my son is but a foolish boy with his head stuck in the clouds."

"On the contrary, sir, his young mind mirrors my own. If you mock him for having his head stuck in the clouds, then I wonder what insult you would proffer me for having mine locked in the stars. Choose your reply wisely."

"Please forgive me, Your Majesty, I did not intend any harm. When the final day comes, a man's character will be judged by God alone."

The King stepped to within inches of Henry Senior's face,

"Well, as this is now my house, I shall be the one to judge your future within it."

The King strode away, leaving Henry's father to consider his future too – for if he had to leave Audley End, he would lose his influence over the villagers he had worked so hard to subjugate. As that thought sank in, he also realised that the man he considered a nincompoop held power over all things, much like a god, and if He indeed was like a god, then he hoped the King would behave likewise, and be merciful.

And so, it came to pass that Henry's father got on his knees, and prayed for the Royal nincompoop to grant him forgiveness.

CHAPTER 13

WHAT'S GOOD FOR THE GANDER IS NOT ALWAYS GOOD FOR THE GOOSE

*I*nside *'Taylor's The Tailor Shoppe' the following morning Jane was naturally bursting to tell the whole world of her new love – but as the whole world would not fit into the shop, her parents would have to suffice...*

"Oh, mother, Henry is so handsome and has such magical ideas."

Her words brought a mixed reaction – both parents being keen to voice their concerns,

"Both will fade with time," was her father's offering.

"Says one who should know," was her mother's.

Thereby both parents turned on each other and failed to give their daughter the support she needed.

But Jane was untouchable, soaring high amongst the clouds with no intention of returning to earth.

"He makes me laugh!"

That obviously struck a chord with her mother,

"Now that is a special gift indeed. Take a firm hold of him and never let go."

"Do you think I shall be fortunate enough to keep him?"

"Fortune will have nothing to do with it! Consider it a military campaign where your strategy is to take Mr Winstanley prisoner without him realising – a willing prisoner, or he shall slip away. A gifted man with humour is a rare catch indeed."

A loud belch from her husband affirmed his wife's point. Getting up from his chair Mr Taylor turned to see both women staring straight back,

"What have I done?"

His wife leant forward, delivering the coup-de-grace,

"My point exactly…"

An idea had entered the King's head to the benefit of His country, but disastrous to any young lovers.

Yet thankfully, young love has a history of overcoming obstacles or the phrase, 'Love Conquers All' would never have been penned.

Young Henry was learning a simple truth – that if you clearly define your goals, you will always be judged by them. Henry's goal was quite simply to fulfil the noble destiny he believed to be his fate.

The problem was – as yet Henry still had no idea what that destiny could be…

Three men of importance were in Audley End's Great Room that day – the vertical one being the King, in ebullient mood, pacing left to right across the entire length of the table in His eagerness to convey exciting news to the other two – being young Henry, who was seated at the centre of the table, his eyes following the King wherever he went, and seated at the end of the table Henry Senior whose eyes did not, remaining stationary and in a glum passion throughout.

"Gentlemen, I find myself in the enviable position of hosting two gifted Winstanleys of the Henry variety. Such a surfeit of talent has compelled me to send one of you on a voyage of discovery for the duration of five years, enriching your skills so that we may all benefit from such richness."

At this point the King came to a halt, addressing young Henry directly,

"Henry the younger, you are to spend time at my cousin Louis's pleasure in France, witness the marvels created by His countrymen, learn from them, improve upon them, then return to create your own, here in my Kingdom."

Young Henry bowed, finding it hard to contain his excitement, however his King advised caution:

"A word of advice for you though: despite His many strengths, my cousin has a weakness worthy of mention - as The Sun King, He believes it shines from every orifice."

Young Henry proffered a sensible question, "Forgive me for asking but how should I address Him, Your Majesty?"

The reply came swift and clear, "I would suggest 'humbly'. He is King of the most powerful land in the world. Take heed of his Court – Louis was but a child when his father died, so has been raised by servants and is thus very protective of them. Also avoid any talk of religion – Louis kneels before the Catholic Cross."

"Should I be afraid of Him, Your Majesty?"

"He has the means to perform good, but also great harm. I live and breathe because of His generosity, allowing me to remain hidden in His country whilst I was hounded here in my own."

At this point Henry sprang to his feet continually bowing with all the energy of youth,

"Thank you for your advice, I am indebted to Your Majesty's generosity and wisdom."

Henry Senior had been watching emotionless from the shadows throughout, and barely moved as he was addressed by the King, whom he still rated with disdain,

"As for you Henry Senior, I have tidings that I trust will excite you too, if 'excite' is the appropriate word to describe a face so rarely seen in motion."

Young Henry studied his father for any sign of emotion, but detected only the slightest of nods. The King continued in the same manner as before, refusing to show anything other than enthusiasm,

"My dear sir, I have been told that you are a great admirer of a certain Mr Wren who, you may be interested to know, I have commissioned to rebuild many churches of my capital that were destroyed in that Greatest of Fires."

Young Henry rolled his eyes at the mention of Christopher Wren – a name his father held in the greatest esteem, and it was during the rolling of his eyes that Henry noticed his father's sparking into life as the King continued: "You shall support his work in whichever form he feels fit."

Henry watched with fascination as his father rushed at the King, kissing His hand with an embarrassing degree of zest, completely opposed to expectations, as he blubbered, "Your Majesty, I consider Mr. Wren to be both an architect and a gentleman of the highest calibre."

Henry watched in fascination as The King crossed over, and tapped his father on the shoulder in a manner suited to the best of friends, or those delivering approval whilst maintaining complete control,

"Wren and I used to play at Windsor together, but alas, he was always too clever for me. I have made this old friend my Royal Surveyor of Works. I can now announce that in that capacity Mr Wren has promised to arrive here at 20 minutes past the hour, which is any moment …."

The King consulted his pocket watch,

"…now!"

The King span on his heels and young Henry sprang to his feet, as his father trotted to the window to witness a coach and four approaching as if on command.

The King smiled, "Aha, one must admire his precision."

Young Henry could not fail to notice his father's excitement, like a child on Christmas morning glimpsing St Nicholas landing in his fireplace, which had also not gone unnoticed by the King,

"I can see Mr Wren's magic has rubbed off on you too, sir. My gift to your son therefore is this - as 'Clerk of Works', young Henry will prosper a great deal from Mr Wren's direct guidance."

Young Henry finally detected the hint of a smile on his father's face.

King Charles strode out of the room in good spirits leaving both Henry Winstanleys to contemplate their offer, standing alongside each other in awkward silence – each being close in proximity, yet nothing more.

They remained in silence for three more minutes until, unable to contain himself any longer, Henry Senior bolted for the door like a starstruck youth.

From behind the curtain young Henry watched as his father raced to the coach performing a ludicrously enthusiastic bow to the enigmatic character alighting from it.

Christopher Wren, in his early forties and finely dressed, politely acknowledged Henry Senior's exuberance then turned his attention to the House, and for the briefest of moments young Henry believed their eyes had met. Losing his nerve Henry backed away from view and faced a terrible certainty - his father would now turn even more distant, for how could his father's son ever match up to his hero...

Ian R Farr

CHAPTER 14
A CHANGE OF FORTUNE

A*s we all come to realise sooner or later - Life can appear cruel in the way it interferes with even the most meticulous of plans…*

The good folk of Saffron Walden looked up as their clock struck eleven startling the seagull perched on top plus the townsfolk below, who sprang into action scurrying this way and that in whichever direction their busy lives demanded.

But in the tailor's shop all was quiet, still and freezing cold – not necessarily from the temperature as it was seasonably mild outside, but Jane was trembling, nonetheless in despair for the marvellous man she had but recently met, and had now being sent away on a mission to heaven knows where, for heaven knows how long. Comforting herself Jane's fingers met the damaged lapel of her jacket instantly, reminding her of their wonderful fishing trip and everything that came from it.... The immediacy of that memory came as a shock, forcing Jane's final collapse into tears.

Small changes in Jane's behaviour lately, greatly worried her mother – the little less spring in her step, the little less enthusiasm in her dealings with the customers, and the little less sparkle in her eyes betraying the fact, that her mind was lost to the man who had taken her heart…

Having heard only rumours of Henry's mission, Jane needed to know the truth, so without telling her mother, she took the family horse and aimed it West for the two-mile, hard ride to Audley End, where she believed there would be answers to her questions.

However, as she approached the imposing Manor, Jane's resolve began to weaken – in plain view of the front windows and the staff, she steered off the noisy gravel drive favouring the soft silent grass of the verge, and aimed for the servant's quarters.

Perched high in the saddle feeling the rise and fall of her thighs from the breathless beast below, Jane paused for a moment, hoping yet fearing what any news could bring.

The clack of a latch made her jump, and Jane turned to see a bright-faced young man stride through a back door leading from the Manor to a smaller door behind which the banging of a knife and clinking of cutlery announced a kitchen. Oblivious of her as he rushed past, the young man's uniform revealed that he was a member of staff by the official 'Audley End' insignia embroidered thereon, immediately recognised by Jane who could well have been the one who had sewed it on in the first place, "Excuse me," she asked surprising the servant as much as herself at her bold delivery.

Upon noticing her youthful beauty, the young man's manner changed, adopting a relaxed approach flavoured with a smile, "Hello, my beauty."

Despite his friendly manner, Jane chose to maintain a formal approach,

"I have come to see Mr Henry Winstanley."

Her heart began to thump, from the excitement of speaking his name.

"Which one, Madam?" he replied, completely innocently.

Jane hesitated - it was a reasonable question, but her mind was set on a single goal and did not want diverting,

"I would have thought it obvious."

"Well, you see, Mr Winstanley – he of the older persuasion – well, he's around here somewhere, but as to where I just don't know, so I can't tell you."

"And the other one?"

"Well, he has long gone."

"Gone where?"

"Well, France to begin with, I've heard."

Jane was incredulous, "To begin with…?"

"Maybe followed by Germany and Italy over the next five years, but I don't know for sure. Now I must get to my chores, so I shall have to say goodbye."

He quickly turned away yet not so quick as to avoid Jane noticing his irritation as he opened the door and disappeared into a cloud of steam that smelt of the lamb stew it derived from.

Jane froze, only a touch wiser than she was before and more frustrated than ever. Then a man crossed in the distance and for a moment her heart leaped for joy – but only for a moment. His walk was similar to Henry's and his face bore a resemblance too, only much older and devoid of Henry's enthusiasm.

Realising who he must be, Jane raced after him and, despite feeling uncomfortable, called out his name,

"Mr Winstanley sir, please stop."

He stopped and turned but showed little interest in her, "Can I help you, Madam?"

As she drew closer, his resemblance to Henry faded with each step, until she was facing an example of what her lover would look like, if he had lost all interest in life.

"Forgive me, sir but I am wondering what has happened to Henry, I have not seen or heard from him in weeks."

His expression changed from disinterest to quizzical,

"And who might you be, Miss?"

"I am Jane, sir, Jane Taylor from town."

"But what exactly *are* you, Miss?"

"A seamstress, sir – he must have mentioned me.",

His voice said "No", but his eyes revealed he had indeed heard of her, and he proceeded to step closer examining her up and down,

"I am the boy's father."

"I know that, sir."

Mr Winstanley paused for a moment intrigued, vanity gaining the better of him, "Then let me ask, Miss – does he speak well of me?"

"Indeed, sir, he told me he wishes to become a son you can be proud of."

Thus, with his ego fulfilled, Henry's father now moved to end the conversation,

"You are too late; Henry is abroad on orders from the King, for a few years I am told."

At this point the manservant Jack was walking past, and on the mention of his friend, stepped a little closer.

Trying her hardest to remain composed, Jane questioned Henry's father again,

"Why would the King of England send Henry abroad sir? Please tell me what's happening, I beg of you."

The question was put with such innocence any man with half a heart could not refuse to answer – but since he did not own a heart at all the reply was brief,

"It is not for me to divulge the King's considerations to the likes of you."

"But I beg of you, sir, when will my Henry return?"

At this point Mr Winstanley's expression changed again from curiosity to irritability,

"The date of *my* Henry's return, Madam, is a matter for The King and I alone."

Jane matched his irritation startling him, "A King, I thought you despised, sir."

Mr Winstanley's face immediately flushed as red as a turnip, unused to being challenged so openly,

"How dare you infer I have anything other than the greatest admiration for the King."

Jack suddenly interjected, attracting both of their attention, "I reckon he will be gone for five years at least, Miss."

Jane noticed compassion in the young man's face giving her the strength to ask - "You seem to know Henry well – "

"Aye Madam, as well as any man can know an enigma."

"Indeed, he can be puzzling at times."

"Yet, we are good friends."

"Then pray, tell me - is he in good health? I have to confess I love him so dearly I cannot find the appropriate words to express how I feel."

"I have heard nothing yet…"

Mr Winstanley swiftly interrupted, sensing the conversation drifting out of his control, "His health is of no concern of yours, Madam."

"I think it is - I believe I am carrying his child."

Both men stopped and stared – Jack adding a smile and Henry's father faltering - his mind racing before asking,

"Have you mentioned this to anyone else?"

"Only to the both of you, sir."

"Good, then let it remain so."

But Jack could not contain his joy, much to Mr Winstanley's irritation,

"My congratulations, Miss, is Henry aware of the good news?"

"Not yet, I need to write to-"

Henry's father, who had been growing increasingly concerned about the direction this conversation was taking, moved to end it completely with raised voice,

"You shall end this conversation immediately, any correspondence between you and Henry shall go through me. Being his father, it is only right and correct."

As instructed, the conversation ended abruptly, leaving both Jack and Jane hanging mid-sentence until she found the courage to ask,

"Please will you keep safely any letters addressed to me, sir, so I may come and collect them from time to time?"

Henry's father nodded swiftly and replied irritably, "You have my word."

"Thank you, sir, I am most grateful. I can see you are a busy man so I promise not to trouble you any further on the subject."

Henry's father added the coup-de-grace – a task he seemed to relish,

"I am delighted to announce my son's intended betrothal to another."

Jane swallowed hard, barely able to speak but managed to utter, "Betrothal?"

"Yes, I assumed you knew..." he added with a false degree of concern and a cruel smile, "...to a delightful girl of my parish. A young lady of fine breeding, who will look after Henry and is pure enough to bear his children."

Having successfully dealt the blow, Henry's father changed tack, turned about and set course for the stables, leaving a vacuum in his wake.

Despite herself Jane shed a tear, annoyed to be showing weakness to the young stranger standing next to her, "Please forgive me, sir, the news has come as quite a shock."

"A shock to me too, Miss."

"Tell me, how well do you know Henry?"

"Well, enough to know something ain't right. Henry don't act like that. I'll find out what I can, and report back to you, Miss – er...?"

"Jane, sir, Jane Taylor. And you are?"

"Oh, I'm just Jack, Miss, nothing fancy - just Jack."

Jack's promise gave Jane some hope, and she held onto that hope for the duration of her ride home, and as long after that as she could manage. It was not much of a hope, but Jane sensed Jack to be an honest sort and besides, she had no other choice...

Setting foot back in the shop at the end of a sad ride, Jane was inundated with questions, almost entirely from her mother. She didn't mind though, as it provided a chance to talk about Henry and in so doing, provided a modicum of relief, as if he was somewhere close by…

…Her father, who until that point had been lost in his thoughts sewing buttons, suddenly came alive interjecting,

"I knew you couldn't trust that one."

"Please do not say that, father, you told me that you liked Henry."

"Like and Trust are two separate things, my love."

"Well, you shall both have to get used to it, because you'll be seeing a lot more of Henry. I am carrying his child."

Her father responded first, dropping a stitch and his spectacles along with it. Yet having already experienced the trials of pregnancy, Jane's mother was wise to it and needed convincing,

"What makes you so sure?"

"My feelings, mother."

"Describe them," she asked out of disbelief as much as concern.

"I awake feeling sick, and have grown prone to weeping at the slightest thing."

Feeling out of his depth, her father decided that women were best left to deal with such matters, and beat a swift retreat to the safety of his buttons.

CHAPTER 15
DESPAIR

Three weeks previously, young Henry had been excited, as well he might have been, for he was about to embark on a great adventure, arranged by the King of Britain to expand its horizons, and in so doing, make Great Britain even greater – for as the King once stated, 'Britain's greatest strength lay in her Innovation', and being an innovator himself Henry Winstanley was the ideal choice to serve his country in such a manner...

In Henry's young mind all thoughts of Jane were replaced by the excitement of the opportunities that lay ahead – the first steps of a journey to prove his worth, and maybe even discover his fate.

He was standing atop Dover's harbour wall on the South-Eastern point of England, being battered by a strong Westerly that sucked up waves from the world's busiest channel to spit at the harbour wall. He was shivering madly, his ears ached and his eyes wept but Henry was happy - for what young Englishman would not be, having had his talent recognised and rewarded by his King.

As Henry struggled to remain vertical against a wind strong enough to force the boats below to lay on their sides like a line of ballerinas, one mast in particular caught his attention - standing upright against the wind, proudly displaying the identity of the World's greatest seafaring nation to which it belonged.

Henry walked up its gangplank underneath the bristling Red Ensign to board the 'Lady Elaine' on the first leg of his journey heading for France, with whom England at that time was fortunately at peace, due in no small part to the King of England being cousin to the young King of France, barely eight years younger than He.

For the moment the weather held fair, but dark clouds lining the horizon hinted at the dangers ahead. Despite the relative calm of the sheltered harbour once his ship moved for the entrance, Henry gripped onto whatever he could find to stay upright, being unaccustomed to being aboard boats at sea. Once the 'Lady Elaine' slid out of the protective harbour wall it immediately pitched and rolled upon meeting the waves of the unharnessed sea, sending Henry skidding across the deck in the fashionable yet impractical shoes Jane's mother had so recently provided for his promised meeting with the King. Inevitably Henry crashed to his knees sending the crew into fits of laughter, not from any ill will, but from remembering their own initiation to the ways of the sea years before. As they laughed watching him sliding across the deck, Henry's good humour made them warm to him all the more until he let out a cry and tumbled over the hatch disappearing below.

In the abyss below deck, Henry grabbed the opportunity of an unoccupied seat amongst the goodhearted passengers, falling off many times before managing to settle.

A passenger (whom I shall call Mr Helpful) said helpfully,

"Worry not, young sir, you shall soon get used to it."

But a bitter passenger with a bitter face (whom I shall call Mr Bitter) added a touch of bitterness, "Aye, but your stomach may not."

Henry tried to keep smiling throughout, "You may well laugh but I have only set foot on a boat once before - to cross the River Cam. I'm sadly lacking in the ways of the sea. Does it always heave so?"

Mr Bitter only offered a grunt by way of a reply, but Mr Helpful added,

"Aye sir it does, but I look at it like this – she is like a lady filling her lungs in order to carry her load."

"Then tell me, sir – does she ever grow angry?"

"Aye, she has a passion like any lady and shows mercy to no man."

"Then I wish her well and pray she grows content for my stomach is not a happy companion at the moment."

Mr Helpful got to his feet, heading for the steps, "I always find it helps to go on deck and fix my eyes on the horizon."

Henry followed suit, heading for the steps, "My thanks, sir, my stomach and I are indebted to you."

Arriving back on the rolling deck, Henry peered through the slanting rain and could just make out a grey horizontal line in the distance.

"My goodness – is that France? It cannot be so close already?"

A strikingly handsome man, blond of hair and blue of eyes showed his experience by striding forward with ease, defying the rolling deck,

"That may be, young pojke, but there will be many a beast prowling down below between us and France."

"I suppose you are right, sir."

"I know I am right, sir. I have a terrible tale to tell that I would wish on no man, be they my closest friend or my worst enemy - unless they want to hear it that is."

Henry was as a schoolboy again, with a mind as open as his eyes,

"Please do tell, sir."

"Let this serve as a lesson, young pojke - as the sun began to rise that morning, we set sail leaving the shelter of our beautiful fjord, and as the sun spread colour across the world we passed over the deepest part, but when it rose to the highest of heights so did the Kraken heaving from the depths to feast."

"Really? What sort of beast is that?"

"A beast from Hell, as none other. It lifted its skirt and I swear I saw a dozen or more of its arms emerge."

"A dozen arms, you say? But that is monstrous."

"Monstrous describes it well, young pojke, for it *was* a monster. Before I had the mind to pray, its arms were all about our ship squeezing until our timbers cracked, then dragged any sailor clinging on for dear life into its beak."

"A beak? But a beak is for birds, sir."

"Aye, and for this beast too. I saw it rip open stomachs and drag out the entrails, look at my throat -"

He parted his beard to reveal a huge deep cut that he opened to reveal his teeth behind.

"My God, sir, you are lucky to be alive."

"Lucky I may be, but I relive that horror every night."

The man growled and turned away leaving Henry gasping.

A hand slapped onto his shoulder making him jump - but it was a friendly hand and therefore a comforting slap,

"Do not listen to him, young sir, sailors are well known for telling tall tales."

"Thank you, sir. I am only bound for France - I believe I can see it ahead."

"Aye, there it is - as beautiful a place as any I have seen."

"And what of the people, sir? We have been at war enough times for them to hate us?"

Mr Bitter appeared, no doubt joining them to spread bitterness, having had no success below deck, "They are a hideous race. They smell on account of what they eat – you cannot trust a man who chews the snails on your wall, and sucks the frogs from your pond. They are beasts, not human. They hate us as much as I hate them."

Henry chose to soften the mood, "I find if you can make a man see the jollity in things, then you have met the child. Win over the child and you can make friends with the man."

"That is a pleasant philosophy from one so young," said Mr Helpful with a smile.

Mr Bitter had the final word, leaving a bitter taste in the mouth for them all to chew on, "Ha, he is only being pleasant because he is so young. Wait a few years until life has taken its toll, then see what he has to say."

A blinding flash heralded the start of a sudden storm, shooting fingers of light across the sky firing Henry's curiosity,

"My father would no doubt say this is God's way of shining a light into the dark, but don't you think it strange how, despite the intense light, we can no longer see France, as if we have been blinded by the light?"

Henry's arm was suddenly gripped tight by Mr Bitter with an unusually helpful proposal that could save both of them, "Come below deck. I have travelled by sea many times, so I know that those who perish are the ones who stay on deck. The fact I am talking to you now is testament to that."

As Henry watched Mr Bitter falling down the steps, he made his choice, calling out, "Mind your step, sir. I shall see you on shore."

Something made Henry look up and whatever it was he was grateful, for he was able to witness a rare but beautiful phenomenon - from the tips of all the masts, in every direction they pointed, flames shot out like burning torches. He was not the only one who was captivated as all on deck looked up and gasped - some from fear yet the crew seemed to be relieved. Intrigued Henry called out to the ship's officer, "You do not seem to be frightened, sir."

"Aye, it is a good omen, lad. St Elmo will guide us to a safe haven".

As if in confirmation, 40 minutes later a sailor at the bow shouted back that he could see a light through the driving rain, a few degrees to port. Henry rushed forward and had to agree - for despite the rain tearing into his eyes and the spray being kicked up by the bow wave below, Henry could indeed make out a point of light ahead that remained visible despite the storm.

"I see it too! You are right - a light to guide us! Thank the Lord."

No sooner had the words left his lips than Henry regretted them, for to give praise to the Lord was akin to praising his father.

Suddenly sailors all along the deck echoed his words – "Praise be to the Lord! A lighthouse - we shall be safe!"

But the Lord, if there was one, had other ideas - visibility was all but zero when the 'Lady Elaine' was persuaded to alter course, and head for the light placing too much faith in the goodness of Man.

Henry gasped as he spotted the rocks racing towards him below, the head of the doomed Captain twisting back to command the Coxswain, the fear in his eyes, the wheel spinning in desperation, Mr Hopeful kissing his crucifix, the light ahead being not a lighthouse after all, but a bonfire surrounded by figures brandishing weapons racing across the beach heading for the ship, the line of rocks closing in ahead like jagged teeth in some giant's mouth, 'Lady Elaine' gliding innocently onwards as the crewmen on the bow saw what was about to happen and fled for the stern.

With good reason.

'Lady Elaine's mouth opened grotesquely wide receiving a granite tooth ripping it apart.

The impact threw Henry over the bow and into the roaring surf, the pressure of which forced the salty torrent into his mouth, until he could take no more overflowing down his throat and into his ears, rendering him both deaf and dumb.

Yet, despite the loss shortly afterwards, Henry spotted movement - the French smugglers hunting in packs, separating the crew, tearing off their trinkets before ripping them open as if gutting fish, slashing their innards and leaving them to perish wriggling in the surf.

He saw Mr Helpful gripping his crucifix as a shield, until a sword forced its way through his back collecting the crucifix on its way out the front.

He saw a fiery man with fierce teeth brandishing a dagger, splashing through the surf to reach him. Then, as incredulous as it sounds, Mr Bitter intervened to save Henry's life by putting his own at risk – not being a natural fighter, Mr Bitter did what he could, wrapping his arms around the fiery man, disabling him as he screamed at Henry to flee. Henry did his best, struggling to move in the sand, losing grip in his fancy shoes, tumbling back into the surf and witnessed, despite the rain and constant barrage of sea, a moment that would remind him of the good in man, much better than any sermon, for despite having no skill in the art of fighting, Mr Bitter had found something worth fighting for, and struggled bitterly with the fiery man until his energy was spent. As Henry watched mesmerised all hate left Mr Bitter's face, and he seemed to grow younger as he entered the gates of Heaven. Left with nobody else to save him, Henry resigned his fate to the mob, hoping that at least they might be merciful.

A hand clasped his throat and, despite choking under the surf, Henry was aware of the crushing weight of the fiery man falling to his knees on top of him tugging at his necklace until the gold case came loose. Suddenly the fiery man stopped – his mouth opening as wide as his eyes upon coming face to face with the King of England. The man's face revealed his dilemma – the narrowing of the eyes indicated the brain was deciding whether it was worth risking its life to take another's – one who was obviously important enough to send him to the Guillotine. The man came to a decision and grunted, releasing a stench of garlic, ripped off the necklace anyway, and pocketed the pocket watch.

The last thing Henry saw that day was the smuggler's blade spinning 180 degrees sending its hilt crashing into his face – then black…

Meanwhile, back at Audley End, Mr Winstanley was leading his hero, Christopher Wren towards Henry's workshop with the aim of belittling his son.

"Mr Wren, could I indulge you for a moment on the subject of my son – a foolish boy, but well-meaning nonetheless."

"Ah yes, I have heard he is an amusing engineer."

"Amusing yes, but nothing more. It will only take a few minutes of your time, sir, but will stop him wasting any more of his on worthless schemes."

Christopher Wren, with his mind occupied on the renovations he had been charged with completing, reluctantly accepted. However, once the curtains were pulled open flooding Henry's workshop with light, Mr

Wren's interest became captivated by all the devices glistening back – Henry's mechanical objects in their various stages of completion.

Christopher Wren sighed heavily, sending the wrong signal to Henry's father who dutifully bowed his head, "Your time is precious, sir - I should not have bothered you on such a trivial matter."

But there was no need for apologies – Wren was lost in intrigue as he uncovered more and more of Henry's devices,

"Do all of these works belong to this young mind?"

Mr Winstanley nodded with embarrassment,

"I'm afraid he is young, sir; I fear without a guiding light he will flounder. If you could be that guide…"

But Wren had arrived at the huge, covered object and, being a Scientist and naturally inquisitive, pulled it away revealing the Solar System in copper form. He let out a sigh, and slowly inspected the contraption, soon discovering the means to set the whole structure into motion. It awoke with a scream of creaking metal, as springs broke free of their tensions, flexing like a newborn child stretches for the first time. It creaked and it groaned as unrestricted metal rubbed against metal setting the whole thing in motion, trembling as leather belts twisted the rods that made the planets rotate. Wren moved back muttering to himself, as Henry's father rushed to apologise,

"My apologies again, sir – nothing but a silly piece of whimsy."

But to his surprise Wren remained fascinated, "That may be – but the scaling is surprisingly accurate…and look…"

Wren pointed to Henry's model of the Earth rotating the Sun, "…remarkable. Your son's belief in the planetary movements mirror my own. I must congratulate you, sir, on raising such an unusually gifted young man."

As Wren left the workshop Henry's father was left speechless - where one would have expected pride in a son's achievements, there was bewilderment.

However, there was a solution at hand - a solution of pressed grapes waiting in his church to soothe him, so Henry senior left his house for the House of God where a bottle stored for the villagers' communion was plundered instead, for the sole benefit of one man - Henry Winstanley senior...

CHAPTER 16

THE CRUEL SEA

Bad news travels faster than good and so it was that a messenger rode through the night at breakneck speed to relay terrible news to Audley End, of a disaster on the Northern coast of France – a shipwreck carrying English passengers, amongst whom was a young man, well known to Audley End Manor, who had set sail on a mission for the King of England…

The messenger that drove his steed at full gallop for Audley End was an honest soul just doing his job, nothing more. He had no idea what the news in his bag contained, but there was one thing of which he was certain - his job depended upon its successful delivery to the recipients concerned, and so it was with much frustration that, after surviving a highwayman's attempt to steal the bag at gunpoint, and leaving the dastard's heart pumping its final dregs into a roadside puddle, his attempts to reach the Manor's front door were hindered by the mass of scaffolding being erected on Christopher Wren's orders for his extensive renovations. Steering his horse around the many ladders, the messenger finally reached the entrance, leaped from his saddle, rushed to the door and with huge relief, handed the note to the doorman – his job complete.

Being slave to the same regulations as any visitor to the Manor, the message was passed from hand to hand – from the doorman's determined grip to a maid's nervous touch, racing with it up many steps before passing it to a young servant eager to impress, who sprinted with it along the entire length of the landing until arriving at Mr Winstanley Senior's room, where he knocked rather too enthusiastically.

Mr Winstanley's hand eventually appeared, opened the letter and after a short pause beckoned the servant to leave, which he did, passing it to one of his colleagues on his way back downstairs.

If Mr Winstanley had felt any emotion that day, he did not show it, reverting to his usual position - on his knees in front of a meagre fire with hands joined in prayer.

Thus, hand by hand the news of the 'Lady Elaine' filtered down the House to the servants' quarters. The shock felt by each gave a good indication as to Henry's popularity with the staff as it was in the lowest level of the House where the highest emotions were felt. The message continued its journey through the kitchen door into the hands of Jack, who upon reading it knew exactly where its journey should end, taking a horse and leaving at speed setting his sights on Saffron Walden, and one place in particular – 'Taylor's the Tailor Shoppe'.

Whilst Jane's mind was focussed on the facts and figures recorded in the accounts, and her mother's attention was fixed upon a customer at the back of the shop, it was no surprise that when Jack turned up, no-one paid him any attention.

Indeed, Jane only looked up once Jack took her hand and placed the letter directly into it with sadness etched on his face. Having only remembered Jack wearing a smile before, Jane braced herself for the worst and as she read the note, Jack noticed her face change: her pupils dilating and her forehead frowning as she digested the message.

Jack was so sad for her, he felt compelled to say something,

"We are all deeply shocked, Miss - Henry was our friend. He was so excited about visiting Europe – he said he had plans but kept them to himself. He said there was only one person in the world he would disclose them to…" Jack concluded with a 'knowing' look to her.

"Why are you talking as if Henry is dead?"

Unable to conjure the appropriate words, Jack nodded at the letter and sadly took his leave.

Being a private person, Jane hid her surprise from the customers, and when she closed the ledger with a deliberate slap nobody noticed, and nor did they notice when she placed the quill neatly alongside, composed herself and reread the letter.

The shop was busy so Mrs Taylor could be forgiven for not noticing her daughter donning a jacket, going into the kitchen and emerging with a bottle of lemonade.

The serenity of the woods exploded with the clatter of birds taking flight from the tree-tops as Jane thundered past beneath, forcing her horse to ignore its fear, racing at breakneck speed into ever denser vegetation, weaving between the trunks of trees, crashing through streams, leaping over fallen branches, sending rabbits bounding, flashing their white tails as a warning to others, until the percussive thud of flank against trunk sent a squirrel scampering to the top of the tree to look down upon the cause of all the commotion - a wounded horse limping away from the broken rider by its side.

Gradually, nature settled back into its rhythms and calm returned to the woods - the tweeting of birds up above, the bubbling of the river down below caressing its rocks, and the gurgle of lemonade flowing from the broken bottle, soaking the skirt of the broken girl whose tears were flowing uninterrupted.

A great tragedy had befallen Miss Taylor that day - her shattered body harbouring a broken hip to complement a broken heart. By coincidence, at that very moment the heart of the town stopped too, as if in sympathy with the situation, for as its clock approached the hour its tenth strike was its last.

Many glanced up to check if there was enough time left to buy the goods on their lists and some, more aware than the rest, noticed the clock's hands had stopped moving, including a new family of gulls searching for a spot to build their nest.

For a while, the townsfolk went about their business as usual, ignorant of the situation until the baker noticed his loaves had scorched and the Taylor family finally realised their daughter was missing.

None of the townsfolk paid any attention to a father transporting his broken daughter along the main street that evening, so confused were they without the reassurance of the clock to govern their day...

However, as tragedy for some is a blessing to others – the family of gulls chose to build their nest on top of the clock tower, content in the instinct they might never be disturbed again...

CHAPTER 17
FRANCE

*A*t this point you could be forgiven for thinking all was lost, but Fate had not finished with young Henry just yet.

The consequences of a near-death experience can vary from deep depression to suicide, but for Henry it brought relief - for it confirmed a belief that he had held since witnessing his brother riding off to certain death in Cromwell's War, and his sister's capitulation to the Plague: the fact he was still alive proved Fate believed he must be worth saving, but for the moment at least, he did not understand why...

As Henry's senses gradually returned it was Sound that arrived first, in the form of laughter – in the mewing of the gulls and the rhythmic swish of sea wash. Vision came next in the form of bodies so recently vertical, now lying horizontal and still.

He spotted Mr Helpful's wooden crucifix floating away from his corpse, deserting him, as others lay facing the sky open-mouthed, creating caves for tiny creatures to explore - the nostrils providing tunnels to the lungs whilst mouths, from where words had so recently flowed, acted as passages to the rich pickings of the stomach.

Henry remembered how he had shouted upon spotting the light, of how he gave thanks to God, and how he had hated himself for doing so. Clearer than everything else though, he recalled the light…the light on the beach that had brought such hope but had finally betrayed them all.

A light, that if used responsibly could have saved every soul.

Despite lying alongside the sea for hours with unbearable thirst, not one drop of water had reached his throat, causing such drowsiness Henry kept falling asleep, at risk of being prey to the insects that were devouring those poor neighbouring souls.

The thudding of the waves was eventually joined by another form of thudding - hooves growing increasingly loud. Shortly he would hear, for the first time in his life, the voice of a decent Frenchman, "Ici! Venir ici! M'aider a le deplacer." (Here! Come here! Help me move him.)

Too weak to stand, Henry turned his head and saw only black – black shapes dancing a distance away – the billowing black robes of a black man riding a fine, black steed directly towards him.

The stallion halted so close to Henry, he sensed the sand shift beneath his neck, felt the pebbles bounce off his cheeks, and the blast of the beast's nostrils on his forehead.

The Moor, staring down from above, chose not to dismount, remaining in his elevated position where his words, although spoken in French, would carry more gravitas, "Bienvenue en France monsieur."

(Welcome to France, sir.)

Henry had not the wherewithal to reply, partly due to fatigue, partly due to dehydration but mostly because he didn't understand French. The Moor leaned over, reaching for a goatskin bag, and offered it down to Henry who instinctively raised it to his lips, gulping at the fresh water with gusto, until he could form words once again,

"Thank you, sir."

The Moor flashed a grin, revealing the whiteness of his teeth, clicked fingers that were adorned with rings of silver, and a short man who looked somehow familiar responded immediately, trotting up to an abrupt halt alongside Henry's cheek from whence he remained stock still, as if awaiting a further command.

The Moor continued, "Monsieur Winstanley…is my presumption correct?"

Henry struggled with a swollen tongue but still managed a polite reply,

"Your presumption is most curious because of its accuracy, sir."

Despite being blinded by a sky forcing the Moor into silhouette, as Henry's eyes adapted he began to notice what a dramatic figure this man really was - from the redness of his eyes, the flecks of silvery grey on his tightly curled hair, the many silver rings adorning each finger, the huge black cloak dripping with silver buttons down to the black knee-length boots finished with ornate silver stirrups: - a dandy, Henry concluded.

His judgement proved correct once this fellow opened his mouth to speak, revealing the gold fillings tracing his jawline,

"Allow me to introduce myself: Monsieur Omar Ricard – with a silent 'd'. I have been appointed by my King as your guide."

"Greetings, sir, my King has sent me here also."

At this point Omar Ricard - with a silent 'd'- snapped his fingers at the short man alongside Henry, who produced an embarrassed grin that smelled of garlic. On Omar's command, the short man lifted his cloak and produced the gold pocket watch bearing the profile of the English King.

Omar's smile disappeared as he posed a question,

"This is a matter of life-or-death Monsieur Winstanley - is this your pocket watch?"

"Yes, sir, it was given to me by your King's cousin - my King Charles."

"Aha," the Moor replied. Henry noticed the flash as Omar drew his sword, and heard the swish as its blade came down upon the short man's neck, rendering him somewhat shorter.

Henry looked down and recognised the decapitated head nestling alongside him – the fierce mouth and the wild eyes of the thief.

At once, Henry felt both disgust and relief, for if this Moor had indeed been sent by King Louis to guide him, then he must also have been ordered to protect him, which was immediately confirmed:

"I am instructed to ensure your safe passage. Your King must value your services greatly."

"I believe it is a favour returned."

Keen to relay the news to his paymaster, Omar Ricard encouraged Henry to climb onto his saddle, then thrusting his silver stirrups into its flanks, sent his steed racing back from whence he had come.

As they galloped along the beach Henry sensed his fortunes were changing for the better and thus, it was a happy Henry Winstanley who clasped the dandy's cloak for dear life wishing for a nearby Inn to spend time in after a night of comfort.

Dear Reader - Life has a way of balancing fortune with tragedy, so following that tragic shipwreck Henry's good fortune was in meeting this enigmatic Moor tasked by King Louis to protect him, whilst acting upon Cousin Charles' orders - confirming that both Kings did indeed place some value on his future, and maybe the reason why Fate had kept him alive...

CHAPTER 18
LETTERS OF LOVE

Henry's mission was back on course – a journey of discovery amongst the grand homes of northern France, during which time he was treated as if he were nobility, for when word got around that he was in the pay of the English King, Henry was treated almost as royalty himself, being afforded the most comfortable of lodgings throughout his mission, where he was obliged to write regular updates for his King, and eager to write letters from his heart to Jane...

My dearest Jane,

Please forgive me for not writing sooner but so many things have happened to distract me of late, I fear you will think I have forgotten you, yet nothing could be further from the truth. Having never travelled beyond our shores I have to admit to initially being somewhat nervous of our King's kind offer, but I am discovering that in so doing He has proven His confidence in me, thereby acting more like a father to me than my own.

My first touch of French soil was so nearly my last, as our ship came to grief on the rocks, killing most on board. All who survived were slaughtered by the wreckers who had lured us onto the rocks with a large fire we believed to be a lighthouse. I believe Fate played a hand in my favour as I was the only soul to survive.

I think you will agree that my introduction to France was a disaster, yet it has taught me two things of which I had no opinion before – a lighthouse would have prevented this disaster saving many lives, and secondly, despite this savagery, the French are proving to be the same as the British - good as well as bad. In fact, I am lucky enough to have made the acquaintance of a most extraordinary gentleman, who has been charged as my guide by no less than the King of France himself.

Omar Ricard – with a silent 'd' – is a Moor cloaked in the darkest skin I have ever seen on a man, and chooses to dress himself in a fitting manner, with contrasting jewels. As such he is what I believe to be termed, a 'dandy'. His choice of clothes is flamboyant in style and entirely black, adorned with silver trinkets which, when supplemented by his greying hair, achieves an overall effect that I believe your mother would call exotic, although I suspect it may be too revolutionary for the good folk of Saffron Walden.

The mention of that place brings such pleasant memories of being with you, my dear Jane, and makes me long all the more to be with you again.

Your loving Henry, X

Exposing his innermost sentiments to Jane, in the romantic bloom of candlelight, made Henry believe he was alongside her. The idea brought such a thrill he vowed to continue updating her of his thoughts and actions regularly throughout the trip:

My Dearest Jane,

Today M Omar Ricard – with a silent 'd' – and I began our glorious King's mission by visiting two Grand Houses and their wealthy owners, a day's ride away, even rowing across a moat to gain access to one of them and fighting off guard dogs at the other, but our efforts were proven worthwhile.

I was captivated by one thing in particular, common to both: the small copper engravings depicting views of both Houses in surprising detail. I had never considered them before as a business proposition, but now I realise just how attractive these engravings were to their owners, for they are tangible proofs of wealth. What is more – they are portable proofs of wealth, much more so than something as large as an oil painting.

I will think on this for they may provide an income when I return, which I fear seems a long way off at the moment. However, the very act of writing this makes me feel closer to you, which is where I increasingly long to be.

Yet I have to admit, that happy time feels a lifetime away at the moment.

So, goodbye for now, my love,

Forever yours,

Henry X

The following week Henry waited impatiently for evening to arrive so that he could write to Jane once again,

My dearest Jane,

As I visit more of these great houses I notice, without exception, elaborate frames hanging from the walls containing paintings in oil depicting the house, the owners, the children and sometimes even the dog - yet of more interest are the smaller pictures etched in brass depicting views of the house. They are proving popular with those of wealth and privilege, so I have decided it would benefit me to learn these skills upon my return. I am a decent painter, so it leads me to wonder - is this what Fate has in mind for me after all?

I am determined to acquire these skills upon my return which alas, feels so far off, I fear you shall forget me in the meantime,

Henry X

Dear Jane,

It was whilst visiting one particularly impressive Villa yesterday that I caught sight of an engraver nervously revealing his work to the owner of the house. I declare, he had no need to be scared for he had captured an excellent likeness of the place. The owner was so greatly pleased, I saw a gift being handed to the engraver, the value of which was evident by the size of his smile and depth of his bow. Some wealthy guests arrived that afternoon, and it did not escape my attention that the owner made pains to ensure that each one of them became acquainted with this artwork.

In this way, I believe the owner, in effect, has gained a portable proof of his wealth, as I had intimated before.

I see a great opportunity here,

Henry X

Three weeks later we find Henry at another Inn, writing under the light of another candle, but the message itself remained essentially the same:

My dearest Jane,

I am fortunate to be witnessing properties with gardens of such beauty to take your breath away as they did mine. My chaperone – Monsieur Ricard, with a silent 'd' – is a worthy fellow, undertaking his task with much patience and a great deal of knowledge as to the history of these places. I am eternally indebted to His Majesty for allowing me this wonderful opportunity to better myself, and if in so doing, the King is pleased, then I am also pleased. How could it be otherwise, for I am my King's servant, and as such cannot demand more than He.

I cannot wait to relay everything I have seen to you. I shall say again, as if it needed repeating: His Majesty is wise and through His wisdom I am learning so much of what I had never imagined before, as you will discover upon my return,

Henry X

A pure soul does not plot to purposefully harm another, they don't think that way, so can easily be used by those who *do* think that way to perform harm on their behalf - Jack could be forgiven for acting with such compliance, passing each of Henry's letters to his father, for they were all addressed to Audley End, and being too innocent to think of opening each, and discovering the smaller envelope inside addressed 'For Miss Jane Taylor's eyes only'. It also never occurred to Jack that Henry's father would not keep his vow, and collect them for Jane – creating an unnecessarily complicated situation due to one nasty, manipulative mind…

And it was not only the mind that had been active, as Jack was to discover a few days later while following a distraught Miss Worthy wandering the gardens of Audley End asking,

"What ails you, Miss Worthy?"

"I do not know; I am so tired and sick - it happens to me often of late…."

"I have noticed this happen to every family I know of - who is the father?"

Miss Worthy stalled long enough for Jack to prompt another question,

"Is his name by chance Winstanley?"

She did not speak - her nod sufficed.

"Does he have a first name, Miss?"

"Henry."

CHAPTER 19
MEANWHILE…

Back at Saffron Walden, certain things had changed since Henry's departure, and not necessarily for the better: the hands of the clock now remained stuck at 4 and provided residence to a nest of noisy seagull chicks…

The townsfolk still occupied themselves with the usual chores, finding them more difficult without the familiarity of the clock to guide their day. However, most failed to notice the change having taken place at 'Taylor's the Tailor Shoppe': Jane was rarely seen, for now needing a stick to walk she chose to spend every day out of the public gaze at the back of the shop, tending the paperwork. As an invalid she had little choice, and moreover with Henry now gone she had no incentive.

This is how we find her today as her mother dished out motherly advice,

"You are wasting opportunities with your head buried in so much paper. Look up and see what you are missing - the street is full of eligible men," at which point she walked over and sat alongside her daughter to either give comfort or press home her point (it was difficult to determine which.)

"Henry is no longer a resident of this world, but as you still are, it is time to consider your situation."

"My situation is quite clearly on the decline, mother."

"Practicality is the issue of the day. I have always said that Ben Drury, the farmer's son is a hard worker. He has had his eye on you for a long time now."

"Yes, mother, but his eye rested upon a woman who could walk and dance and bear children, none of which are available to me now. Besides, I have no interest in the matter."

"Be practical and take the buggy to Homer's Farm tomorrow, I shall get word to young Master Drury that you are coming. Who knows – maybe he will own that farm one day, but remember, - Ben is the best of what is left. From this day forth you must cut your cloth accordingly. I shall pick a skirt for the occasion."

Jane, who had heard enough of this, got to her feet with difficulty… which did not go unnoticed by her mother, "…a skirt with a wide girth."

CHAPTER 20
THE PALACE

D*earest Jane,*
As I was writing to you today Mr Ricard - with a silent 'd' - brought the most wonderful news, turning this day into one I shall never forget, and surely of great import for the both of us -"

Yours,

X

The observant amongst you will have noticed that Henry had signed off his letter on a positive note, revealing how much Jane was still occupying his mind...

A tap on the hotel bedroom door had interrupted Henry's quill in its dance across the page. The culprit, Mr. Ricard, had popped his head around the door and with a mischievous grin, beckoned Henry to join him...

A day of Sun and brilliance began just as described – a hot sunny day reflecting a King's obsession with magnificence - King Louis' Palace.

In his role as chaperone, Monsieur Ricard had arranged the carriage that had transported them both to the Palace of Versailles, where Henry was introduced to his equivalent French Clerk of Works. Under his guidance Henry was led on a guided tour of the place through doors so magnificent, one could be forgiven for thinking you were entering the gates of Heaven - and in many ways Henry believed he was, from the exuberance of splendour on offer, noting his observations with ink upon paper - detailing descriptions punctuated with occasional sketches observing the many pillars of gold supporting elaborate ceilings from which decorative walls displayed magnificent paintings, depicting past glories in the hope of yet more to come.

In the midst of such magnificence, Henry struggled to describe them well enough for his King to choose which to put before the finest artisans of England, thereby achieving a superior footing to their French counterparts, allowing England to beat the French artistically, before crushing them militarily.

But Henry's task was made near-impossible by the barrage of fresh attractions that continually appeared as he strolled through the Palace under huge arches from which candelabras dripped bunches of glass grapes in a sparkling atrium as visitors, courtiers and staff alike took advantage of the huge mirrors to admire themselves in reflected glory. He was chiefly attracted to a series of illuminated fountains placed amongst the tableware to dramatic effect, firing his imagination. Reaching for his quill, Henry earnestly committed them to paper to scrutinise later that evening, when he could try and understand their workings - so much had they intrigued him.

"So, Henri, has your opinion of our King changed at all?"

Ricard's interruption snatched Henry out of his reverie needing a few seconds' composure, so deep had he been under its spell, "Did He commission all of this?"

"Of course He did – He is the King after all."

"Then I think France is extremely fortunate to be led by such a great man."

"It takes a truly great man to rule over the greatest power on Earth."

"I agree, and I applaud Him for that."

"Good, then you can applaud Him in person later today."

Henry immediately panicked,

"Today? But I am not prepared…"

"Then prepare yourself, and fast, Henri - He has no patience for fools or timewasters. Follow me through to the gardens and be quick – he is immensely proud of them, and will ask your opinion. Quickly now…"

Bursting through huge doors into brilliant sunlight Henry gasped, convinced he had been rendered deaf and blind by all the magnificence on offer.

Initially Henry felt the relief of ignorance, as if awakening from the deepest of sleeps where ears needed nothing to hear and cheeks were warmed by a sun that blinded all sight – but gradually, as his senses returned and his eyes adapted to the glare, Henry was able to see the enormity of the task before him in chronicling all of the surrounding splendours: the vast ornamental ponds – one the size of a small lake, that were surrounded by decorative gardens as if in a grand map where people wandered along pre-determined routes designed to view the whole concept at its best: – the flowers in full bloom, the bushes adorned with berries, the elaborate statues depicting aquatic creatures with huge pursed lips separated by unusual curves …

Ricard watched in disbelief as Henry stepped over the rim to examine those lips, "I suggest you come back here, Henri – I can hear the music."

But Ricard didn't appreciate the fact Henry had a Scientific mind, and as Science is driven by curiosity, when Henry stepped closer to examine the fish, Ricard was already too late to save him – the music swelled and the inevitable happened – water ejected from every mouth as the sculptures sprang to life in a thunderous climax, blasting Henry off his feet. And now the cascades were making sense of the sculptures - the unusual curves were channelling the flow of water over stone leaves, whilst cherubs blew plumes of water from pursed lips.

It was only when he was persuaded out of the water that Henry heard the distant sounds of music - the hooters, whistles and drums that grew louder accompanying a colourful band of marching clowns approaching like a toy army.

Youthful dancers with painted faces, laughed as they twirled but failed to impress Henry, considering them nought but flamboyant clowns, such that when the most flamboyant clown of all closed in like a spinning top, Henry raised his fists in defence. Ricard raced across in a panic, restraining Henry with a bear hug, pleading with the clown,

"S'il vous plait, pardonnez-lui, Henri ne comprend pas." (Please forgive him, Henry does not understand.)

The main Clown raised his hand, and at once the music stopped dead, bringing the dancers to an abrupt halt like clockwork soldiers whose spring had suddenly sprung.

He then approached so close to Henry he could see the genuine lips behind the painted smile as he spoke,

"Ne vous inclinez-vous pas devant Dieu?" (Do you not bow before God?)

An elaborately dressed Midget, bearing a countenance of great import, raised a toy cone and translated in such a squeaky voice, all notion of importance was lost,

"Do you not bow before God, monsieur?"

Ricard stepped forward, explaining in Henry's ear,

"He believes the King is God's sole representative here on Earth."

Ricard awaited Henry's reply nervously, which soon arrived bold and clear, "I bow only to my King."

The clown interjected, "Par votre accent, je crois qu'il est mon cousin Charles. Il deviendra aussi mon sujet des qu'il foulera le sol francais." (By your accent, I believe you are referring to my cousin Charles. He will also become my subject, once he steps foot on French soil.)

The courtiers exploded into laughter, which only served to antagonise the bewildered Henry as the Clown continued,

"C'est un homme bon...mais une fois qu'il se joindra a mon Eglise Catholique, il deviendra un grand." (He is a good man...but when he joins my Church, he will be a great one.)

The group erupted into laughter again, prompting Ricard to translate,

"He says your King will be a better one for joining the Catholic faith."

"Do you believe He will switch faith?"

"Personally, I do not and privately, neither does my King, but He lives in hope."

"Your King places more faith in religion than I."

The Clown spoke again, but Henry couldn't tell if He was joking behind that painted smile,

"Vous etes tres mouille, expliquez-vous." (You are very wet, explain yourself.)

The troupe fell silent, all eyes targeting Henry, awaiting his response, but as he didn't understand French, Henry remained stock still, lost for words.

Ricard stepped forward performing the role of chaperone,

"He is asking why you are so wet."

Henry could only manage to formulate two words - "Je" and "English."

The Clown laughed, allowing the Midget to laugh which released all the courtiers to join in, leaving Henry utterly bewildered, looking to Ricard for guidance. The voice that spoke next belonged to the Clown,

"Cela explique tout, tous les anglaises sont mouilles."

The midget, obviously enjoying Henry's dilemma, translated gleefully,

"That explains everything, all the English are wet."

Whereupon the Clown interjected, affirming his true identity,

"Il y a une exception. Le cousin Charles est un bon Anglais." (There is one exception however - my Cousin Charles is a good Englishman.)

Henry whispered to Ricard, "Your King is a clown? That explains a great deal."

Ricard was furious indicating Henry hold his tongue. Protocol took precedent again allowing the Clown to laugh first, followed by the Midget before the courtiers all joined in competing to offer the most impressive praise,

"Votre Majeste est tellement drole. Nous avons le privilege d'assister a un esprit aussi incomparable." ("Your Majesty is so droll. We are privileged to witness your unparalleled wit.")

Then, as promised, the King turned to Henry and posed the question,

("Alors, que pensez-vous de mes jardins? Soyex averti que votre response doit etre flatteuse ou je la considererai comme une insulte personnelle.")

"So, what do you think of my gardens – and be warned, your answer must be flattering, or I shall consider it a personal insult."

With all eyes upon his response, including Ricard's, Henry felt flattery was the order of the day,

"Magnificent, Your Majesty."

Yet, they waited for something more, compelling Henry to ask,

"What do I do now?"

Ricard's reply came sure and swift, as if it was obvious,

"Now, you must bow!"

Taking instruction, Henry bowed but went overboard with hands gesticulating too lavishly, amusing the King nonetheless,

"Je vois que vous etes un peacock Sire." ("I see you are a peacock, Sire.")

Henry did not understand but as he hadn't been arrested, he assumed he would not be thrown into a cell just yet.

The courtiers headed for the Palace like a class of giggling children on a school trip. Henry was at a loss, for this clown, a mere six years older than he, was prancing around in a flouncy costume like an idiot - and what made him even more perplexed were the folk dancing behind in the same manner screaming with laughter at every one of his jokes, begging the question -

"Is that clown truly your King?"

"He truly is."

"But He's nought but a flamboyant fool."

To Henry's complete surprise, Ricard took hold of him and shook him hard like a naughty child, "It would be foolish to underestimate Him - He may be young, but he made you look the fool. At least He will remember you. Your King will find this all very amusing when I report back."

Henry found the last remark depressing – his hopes of achieving success did not include being ridiculed, and realised he needed to make amends,

"My apologies, Mr Ricard, that could have gone a great deal better."

"Yet, it could have gone a great deal worse," replied Ricard performing his role perfectly, "Do not trouble yourself Henri – at least He will remember you. Now come and look, I have been instructed to show you His Palace of which He is very proud - it is hard to believe it used to be a hunting lodge, that He had rebuilt into what you see now...."

Ricard led the way back into the Palace, following crowds of courtiers who had joined the clowns following their King like bees to their Queen.

Upon entering a huge room leading to the Royal Bedchamber, Henry gasped - his Quill springing to life like an insect in panic, dancing across page after page in an attempt to describe the many fantastical wonders on such small pieces of paper, and in such a short space of time.

Mechanical toys of ingenious design were being carried to the Royal bedchamber, dazzling Henry to such a degree he had no choice but to laugh out loud,

"What on Earth are all these, Mr Ricard?"

"They are being prepared for The King's amusement."

"So, I was right - He is but a child.",

Mr Ricard summoned all of his considerable patience, "A crown would weigh heavily on any man's head, let alone that of a child. He lay to sleep as a five-year-old boy and awoke as leader of the greatest power on Earth. He took charge of everything – the Law, the Courts, the Military and the Agriculture raising France to where she sits now – the most powerful force in Europe with the greatest Palace of all - in which you are presently standing, and I must say, looking mightily impressed! His Servants took the role of parents, so it is little wonder He holds a great affection for them, and they for Him. He loves to play, He loves to dance, He set up an Academy of Dance, he loved it so. You see, our King was denied a childhood, so forgive Him, Henri - the French understand His need of time to play."

Duly chastised Henry took to pen and paper once again...

My dearest Jane,

Europe has set my mind free, making all the walls my father built to contain me tumble as if made from mere sand and spittle.

These last few days I have seen such wonders to have inspired me to sketch some of my own. Visiting His Palace at Versailles was like walking into a giant box of toys from many a Christmas past. My eyes were as wide as a child's amongst all the fantastical creations that I am sure would have taken your breath away as they did mine: a miniature Royal coach with flags pulled by horses, a balloon that floated before my eyes fashioned from the bladder of a pig, and a contraption that made me laugh and laugh – a robotic duck that when placed in water swam like a duck, and when lifted from the water walked like a duck, and to my surprise it even proceeded to quack like a duck and lay an egg!

Many things are done differently here, some of which have inspired me to commit my own ideas to paper.

And so, dear Jane, maybe, amongst all these fascinations lies the wonder I have been placed upon this Earth to achieve. Fate will reveal it to me one day; I am sure of it. Now that His Majesty has allowed the return of Christmas, I look forward to it immensely because it brings the possibility of being with you, my dearest Jane, and the realisation that despite our brief time together, I have decided to ask if you would spend the rest of your life with me, as my wife.

Your ever hopeful Henry

X

Henry's letter, asking for Jane's hand, travelled approximately 100 miles before landing in those of his father, who made sure it travelled no further, dropping it on top of all of Henry's previous letters, in a box hidden beneath the chamber pot underneath his bed.

CHAPTER 21
CONSEQUENCES

*A*t this point I should point out that God may have gifted Farmer's son Benjamin Drury with an impressive physique but that was where His generosity had ended, for young Benjamin's strength lay in the muscular rather than the intellectual side of his character....

If this were a tale of 'faces' then Jane's could not have been more different to that of Ben Drury the farmer's son, whose grin was as wide as his horse, as he watched Jane approaching grim-faced in her cart through his farm gates, coming to a skidding halt only feet away from his knees, kicking dust into his face, making his grin appear all the whiter for it.

Jane fired the first salvo delivered as a statement, as opposed to a question,

"I have been instructed to meet you. I trust I am not late."

The voice that replied bore no malice or sarcasm, just honesty, delivered with as deep a tone as one would expect from such a large manly chest as God had gifted Benjamin Drury,

"Your Ma told me to expect you. I have to say it's been a long time coming. I have had to turn down two other girls. You see how lucky you are."

Jane's face remained fixed, "Are you are aware of my predicament?"

"What do you mean?" Ben asked, his large eyebrows raised in the manner that delighted so many of the local girls.

Jane's face remained stern, "My mother has not told you then?"

Ben's impressive eyebrows remained raised, "Told me what, my beauty?"

"Get in."

The day was both sunny and hot, reminding Jane of her day with Henry, which now seemed a lifetime ago, indeed as if it had never happened at all.

Without ever glancing at Ben, Jane forced her cart across the fields at high speed, challenging Fate to destroy her. Being supremely self-confident, Ben was amused, and his grin only widened.

"If your intention is to impress me, then you already have, sweetheart. I wager you take a lot of controlling, but I'm up to the challenge."

Ben made to hold her hand, but was swiftly pushed away. Unused to rejection by any woman, Ben's passions came to the boil stoked by the hypnotic sway of her bosom as the cart jolted and bounced. Determined to have his way with her, Ben set the wheels into motion reaching for her cleavage, but was thwarted by a hearty slap and a deflecting question,

"Tell me, Ben, what you think of those birds."

Ben was predictably thrown, "Birds is birds, Miss Jane."

His answer was as simple as Jane had expected, yet without the malice she had also expected, so she decided to play a little with him in case there was anything surprising worth discovering,

"Birds are indeed birds, as you point out, Ben, but I am curious what you think of them."

"I think they would taste good in one of my Ma's pies."

"I see, and those trees, Benjamin, what do you make of them?"

"Trees is trees, sweetheart. They would look good on my fire – why are you asking such odd things of me?"

Jane brought the cart to a sudden halt, "To see what kind of a man you are."

He suddenly flushed, rising to the challenge, "I will show you what kind of man I am, sweetheart."

Ben lunged forward, fired by lust, smothering her face with his mouth as his hands tugged down her shirt exposing her shoulders to his tongue. Ripping his own shirt off, Ben caught Jane's eyes registering his form just before he found her breasts and devoured them. Jane surprised herself, responding to his pursuit, her own passions rising in need of fulfilment until Ben stuffed his hand up her skirt and stopped as abruptly as he had started, his face confused as her condition dawned upon him, retracting his tongue back to where it belonged.

Jane stared, equally startled, witnessing his descent from man to confused little boy, muttering until he could take no more, then leapt off the cart, and bolted for the safety of the woods.

For twenty minutes or so Jane lay exactly as Ben had left her, spread on her back in the cart, legs open wide, her womanhood fully exposed.

An understanding dawned upon Jane, that she may never again feel the fulfilment of physical passion, or indeed ever experience the pain of childbirth. So, with the tightening of each button and the tugging of each crease she was, in effect closing shop, and by the rubbing away of each tear Jane was, in effect, erasing her path to womanhood.

Accepting her fate, Jane tugged on the reins, turned the cart around, and headed for home.

CHAPTER 22
HEADING HOME

Five years is a long time, and many things will change, but things that matter tend to stay as they are, for as Henry began heading home, so did Jane, into her mother's arms, and was promptly plunged into a wedding dress.

"Unhand me, mother, I do not wish to wear it."

"Your wishes are no longer relevant, my dear – this has cost your father and I a pretty penny. Every girl's eye will be on you, and each shall be green with envy. Benjamin Drury may not have been worth the pay-off but we must face facts - he is the pick of what remains of the crop, and it is my job to ensure you look ripe for the picking. Our order books shall be full for months."

"But I hate the colour, mother."

"Stop being silly, there is only one colour for a bride."

"But I am not a virgin."

"Who is these days?"

"It is the wrong size."

"Rubbish – lest you have forgotten, I know your size better than anyone – I felt you growing inside of me throughout the unbearable heat of that summer, the sadness of the autumn and that most wearisome of winters. Allow me the joy of seeing you paraded through town in a dress of my creation, knowing all the girls' eyes will be upon you, and all of them coloured green from losing any claim to young Master Benjamin. Being this town's tailors of choice, we shall prosper."

"But I have no wish to live with that man."

"It is time you faced facts, my dear - the deal is done."

The finality of those words hit their target as surely as an arrow through the heart.

Mr Taylor had watched silently throughout, observing the light draining from his daughter's face until he could bear it no longer, and leant forward in an attempt to provide some reason,

"Your mother and I cannot run this business forever, my dear. This shop may carry our family name, but you will be the one left holding the reins."

Jane's mother interjected, determined to make her point,

"I am sorry, Jane, but a union with Mr. Winstanley was never going to be on the cards. He has been gone five years now, and will have made the acquaintance of many other girls – I am told the ladies of France are dainty and pretty, the girls of Germany full and buxom, and those of Italy as dark as the Devil, and just as intoxicating."

"Thank you for your guidance, mother, but I shall let my heart bear that responsibility alone."

Her father concluded the conversation, "Let's face it, my dear, Henry was never going to marry a seamstress."

Clumsy words intended as caring, struck deeper than any her mother could have mustered, leaving Jane speechless.

Yet, a glimmer of hope still glowed within her like the embers of a fire, and like any fire, all it needed was a spark…

The white cliffs of Dover were a welcome sight, greeting Henry like a row of teeth set in a permanent smile.

Just over 60 miles to the West, a coach and four was heading past Greenwich Docks and its forest of ships without attracting anyone's attention – and indeed why would they, for many such 'upstart four-wheeled tortoises*' as they were mockingly branded, passed that way each

week carrying paying passengers from the newly-arrived sailing ships to the commercial centres of London.

What made this particular 'upstart tortoise' of relevance to our story was the passenger seated therein…

Henry sat with his head out of the window, taking in the sights passing within yards of his face – some comfortingly familiar from five years ago, and others altogether new.

One of the 'altogether new ones' captured Henry's attention in the distance as his coach passed Greenwich Observatory, still under construction high upon a hill beyond the River Thames,

"Pray, what is the curiosity being built over there?"

Henry's companion was the same colourless Clerk who had announced his adventure five years and a lifetime ago,

"I am told it shall be an Observatory for the study of heavenly bodies. The King wishes to solve the questions of longitude and latitude to aid navigation…and more besides…"

"How more?"

"I have been told the intention is to determine Time itself, so that the whole world shall calculate the time of day from that very point."

Henry was immediately captivated like a moth drawn to light,

"My goodness, that is fantastical! Who is responsible for creating such a place?"

"Mr Robert Hooke. I am told he studies things that are too small for the eye to see. A waste of time bordering on stupidity, if you ask me."

Henry smiled, "Jolly good, I thought you were going to mention another name that continually plagues me."

"However, Mr Hooke has not the capability to complete it on his own therefore there had to be a collaboration with the only man who can impress The King in that way."

"Who so, dare I ask?"

"Why, the genius Christopher Wren, of course."

Henry received the news with less enthusiasm than it deserved, commenting resignedly, "Ah, but of course – of all the geniuses it would have to be Mr Wren."

Continuing its travel Westwards, the coach passed a spectacular building under construction on Ludgate Hill, much larger than the Observatory,

"And that, young sir, is being rebuilt like the phoenix from the ashes to become the most magnificent cathedral in the land, maybe even the whole of Europe. It will be named after Saint Paul as before. I am told the architect, in his passion for astronomy, is designing enough space to install a huge telescope in one of the towers in order to observe the heavens, as his creation comes to life. A waste of effort if you were to ask me."

Henry gazed at the truly impressive construction being created by the hands of many gifted craftsmen milling around the site.

"Mr Wren is responsible again no doubt?"

"Of course, I am told he is a genius of both mathematics and science. Truly a man of wonder."

Henry resented the Clerk's enthusiasm, replying under his breath, "Aye, wonder indeed," intended for his ears only whilst a thought sprang into mind, intended for his mind only, "Is there no end to this infuriating man's 'wonderfulness'?"

Over the course of the day, as the willing steeds dragged his coach Northwards, Henry prepared himself for the reunion yet to come, reasoning that a meeting of father and son should surely be an occasion for gaiety.

Yet, gaiety was the last thought on his mind, and he prepared himself for the worst...

The following day, upon alighting from the carriage at Audley End, Henry could not help noticing the many wooden structures that had sprung up around the House since he left, supporting platforms along which men were practising their crafts using the varied tools of their trades, either spreading cement onto bricks, sawing at wood or replacing glass – a hive of activity obviously following a pre-determined plan.

Henry knew there could be only one person responsible for such a plan and it pained him to admit it.

Racing through the House in search of Jack or his father, all Henry could see were workmen he did not recognise going about their business until he skidded to a halt, staring at the wall dead ahead, aghast – for in this most prime of spots, mounted high upon the wall of the Great Room, was the culprit – the man responsible for all the restorations and the hero of his father – Sir Christopher Wren, captured in oils for all to admire - his knighthood confirmed in brass at the bottom of the frame.

As if predetermined, a servant rushed past skidding to a halt upon noticing Henry's astonishment. Putting two and two together remarkably quickly, he uttered,

"Your father insisted he took pride of place just there, so that upon entering the Great Room his presence would be felt by all."

"Indeed," replied Henry, unwilling to enter into a conversation whilst under the influence of such anger, so headed for the Servant Quarters to find Jack – but alas, to his discomfort Henry failed to recognise anyone he passed and, more alarmingly, none of them recognised him.

Feeling like a trespasser in his own place of work, Henry searched the once-familiar passages until he heard someone calling Jack by name. Rounding a corner, he came upon a servant organising a group of maids. He was obviously the one in charge, but the maids were acting unusually familiar with him, as they would with a friend, and the more he studied the servant, the more he noticed familiar traits – the pinched nose, the dimpled chin and the missing lobe of his left ear, sliced off by a farmer's knife many years ago for poaching a pheasant on his land – a case of the punishment far outweighing the crime. But, as he could only see the man in profile, Henry needed to risk a greeting to confirm the fellow's identity,

"Hello, Jack," Henry whispered, not wanting to startle the man in case he had been wrongly identified.

The man continued to address the servants as if he hadn't heard, finally turning after a prompt from one of them. It was only when he fully turned to face Henry, that his identity was confirmed: yes, this was indeed

Jack, but fate had been unkind - whatever had happened to Jack over the last five years, had sucked some of the life out of him.

Jack's face contorted as it tried to identify the man addressing him, then Henry noticed a spark of recognition as if the brain had only just received his image. A smile formed immediately afterwards, clearing his face of grief, plus whatever sadness had overtaken him, and five years fell away in an instant.

"Henry? My goodness, I knew you would return sooner or later, but I must admit I thought it would be sooner than this. Your return from the dead shall bring this old house back to life."

"Thank you, my dear friend, but why did you say, 'from the dead'?"

"How could we have known otherwise?"

"But the King made it clear my mission was for five years. He has duly kept His word."

"And thank the Lord for that - it is so very good to see you, Henry. Does it please you to be back?"

"I must confess my eyes have been opened to so many wonders in Europe, my mind is still bedazzled by them. I have the King to thank for that, as I expressed to Jane in all my letters."

"Then I expect she must feel good for you."

"Expect? Have you not seen her then?"

"Why would I? I have not heard from you, so why would I have contacted her? As far as I know, in all the excitement you must have forgotten her."

"What? She was on my mind day and night, as I expressed in the letters I sent to her, through my father."

Jack's expression changed from innocent to suspicious in a heartbeat,

"Which letters?"

Henry paused for a moment, assimilating the news, hoping it was not as he was now suspecting, "The letters I sent to Jane through my father almost every week – over two hundred in all."

Jack gazed back blankly, "I have not spoken with your father since you left, but I have noticed he moves a little slower these days."

"Why is that?"

"Oh, that is obvious, don't you think?"

"It might well be, Jack, if I knew what the Devil you were talking about!"

The final words were delivered with a taste of sarcasm that surprised Jack, stalling him for a moment, "Becoming a father at his age would exhaust any man, let alone one who has grown fat on an excess of port and cheese since you left, resulting in him sleeping most days into the middle of the afternoon when he should be out working for the church."

"Becoming a father? My poor mother must be exhausted."

"Much has happened since you departed, Henry – your mother did not provide the womb, but a woman so gaunt, it was a miracle she managed to produce anything other than a worm."

"Aha - I believe I know who you mean, but tell me something of more importance - what has happened to Jane? Please do not say she has married, for I could not bear it, although I have to confess, I cannot blame her if she knew nothing of the letters."

"You have but 45 minutes to get to Saffron Walden before she is wed and out of your reach forever…"

A thunderclap brought Henry to his senses and pulling a cover over the cart he exploited Toby's eagerness like a man possessed, flapping at the reins, willing the young steed to pull him ever faster bouncing along the increasingly familiar route back to Saffron Walden and in particular, Saint Mary's Church.

The once-noisy nest atop the Clock Tower was presently host to a family of blackbirds - the seagulls having deserted it two years ago to the delight of the townsfolk who preferred trading their squawks for the fluted tunes of the new tenants.

Expectant guests were milling down below around the Church grounds – ladies complimenting each other's posies, men planning their post-service beers, and girls still bemoaning the loss of their idol, Ben, the farmer's son, to a 'Plain Jane' who did not deserve him.

Nearing the church Henry's heart dropped at the sight of twenty or so families ahead, all dressed in their 'Sunday Best' in anticipation of the wedding yet to come: farmers looking awkward in their pressed suits, egged on by wives making the most of the occasion engulfed in decorative dresses, no doubt provided by 'Taylor's the Tailor Shoppe' whose owners Henry spotted through their window fussing around inside, as if in a state of panic, as he passed. Beyond all the fussing, rushing and pushing, something in front of the church was moving erratically that on any normal day would have drawn only mild attention, but this being a wedding day, drew no attention at all: a bush that shook erratically by a force that became clear once the young perpetrator's head bobbed up, encircled by a necklace of rosary beads with an aquiline nose wrapped in olive skin, separating two dark brown eyes that were scanning the proceedings with all the keenness of a spy....

Henry pressed Toby onwards ever faster, refusing to slow down through the busy market town, passing the prettily-dressed girls all weeping for the loss of their most eligible bachelor – Ben Drury – into the arms of a woman unworthy of his charms - '...why would he choose such an ugly, plain Jane', he overheard them say. Their grief was so all-consuming as to miss Henry on his cart charging past them along Main Street, heading for Homers Farm beyond town, pulled by his enthusiastic young horse Toby – a beast with nothing on its mind beyond the apple reward waiting at the end. Dust that had kicked up from the cart, floated above the High Street for a second or two before gravity dragged it back down coating both of the girls, giving them another reason to weep, plus the young dark-haired man with the aquiline nose, who will come to play a major role in this tale:

Introducing Mr Jacob Wiley, bearing a forename that suits the financial type he was so keen to join, and a surname indicating how far he would go to achieve it.

Until now Wiley's modest clothes had served what was required of them, but having just overheard news of 'an ugly plain Jane' about to wed a young man of property, he had need of a new outfit for the occasion ...

As Jane's parents panicked over finalising her dress, they failed to notice Jacob Wiley surreptitiously climbing over their back wall and entering the shop via a rear window.

Once inside, the sounds of the panicking parents provided sufficient cover for Wiley to go about his business unhindered – stealing a tie from the displays plus a jacket from its hanger before racing away, and in his haste, disturbing an alert dog that barked with such gusto it caused Wiley to panic, making a small mistake that would create a big impression later – for whilst leaping the wall and descending the other side, he got caught on a protruding nail, nicking his jacket, snapping his necklace and slicing his cheek, thus branding himself with a permanent sign of his true nature.

Meanwhile, whilst racing for Homers Farm at the edge of town, Henry spotted a carriage and four slowly approaching, carrying a bored Benjamin Drury in the rear, looking awkward in his wedding suit, seated behind his stern-faced parents teasing the reins – obviously a carriage full to the brim with indifference.

Inevitably, Henry's nerves got the better of him as troubling thoughts began to appear - for if the girls were right and his true love had indeed turned into 'an ugly plain Jane', did that mean she had lost all her hair, plus maybe three or four teeth and grown dumpy as a pudding - for five years would surely take its toll on Venus herself?

But that only led to another troubling thought – would Jane still find *him* attractive after five years, for he had suffered a degree of reshaping – the gain of weight around his neck being more than matched by the increased circumference of his girth.

Thus, Henry decided to curb young Toby's eagerness, slow down a tad and consider his options – whether to wed, or not to wed...

Ben appeared rounding the corner ahead, and within seconds the two carriages met in a cloud of dust, surprising both Ben and his parents in equal measure, all secretly hoping for a reason to delay the proceedings.

As if in answer to their prayers, Henry leapt off his cart to confront the baffled Ben, extracted his purse and gestured for Ben to help himself to the wealth within (which to a farming family used to trading in cash, was akin to holding a red rag to a Bull).

The purse was swiftly emptied and the Drury family, who had approached so slowly, now span 180 degrees and beat a hasty retreat, satisfied with their easy profit.

In the town, as time seemed to be passing slowly with no movement from their clock, the good folk of Saffron Walden waited respectfully for the wedding to begin and bring another excuse to lead a merry dance of drinking, gossiping and canoodling. Nobody paid any attention to the dark-haired opportunist, with the aquiline nose wearing his rosary necklace into church, then sliding onto a pew at the back, waiting for an opportunity worth exploiting…

A murmur was building beyond the church gates resulting in much turning of heads to witness a wedding dress of considerable beauty encasing a bride hobbling along the street on crutches as glum as if she were attending an exhibition of pitchforks.

Meanwhile, inside the church, non-believer Henry was waiting with irritated compliance and some nervousness, as the piano swelled to the tune's climax to introduce the bride. Performing a degree of decorum, for the benefit of politeness Henry nevertheless was intrigued by a sound he had not expected – the click-clack of a wooden staff approaching down the aisle, releasing a smattering of polite whisperings from the respectful members of the congregation, and a few girly giggles from those who were not.

Determined not to comply with falsity, Henry remained stock still maintaining a steely concentration on the stone floor at his feet, glancing to neither left nor right, refusing to indulge in a ceremony celebrating a greater power he held no belief in.

Moreover, although he wouldn't want to admit it - Henry was growing more nervous by the second…

The click-clacks grew louder then slowed, and as he stared at his feet an elegant white shoe came into view stopping directly alongside him, containing a foot he recognised as the one he had kissed five years and a lifetime ago. The wooden crutch arrived abruptly afterwards, accompanied by the final 'clack'.

The realisation that Jane was finally alongside him after so many years, started Henry's heart beating so hard, he was sure she could hear it. Daring to glance down at her, he noticed, with great relief, that her hair had remained the same, and the steady pulse on her neck quickened with the touch of his hand on her arm. To the surprise of those close to him, Henry bent low to take in her scent, revelling in the aroma causing him to sigh. Jane immediately flinched, unused to such a delicate sound coming from a man, and dared to glance up at the one now standing beside her.

Henry stared as open-eyed as a child – he could not help it; her eyes were just as he remembered, as were her cheeks and the mouth he had kissed by the riverbank all that time ago.

As they gazed at one another, they bonded again without need for words.

The words that did come were from Father Parsons asking for the ring. Henry duly complied, presenting a most unusual specimen whose bulk had a reason, for it represented a metal fly perched on top. Once Henry's finger touched the fly a golden trout sprang forth catching it in its mouth. Jane released a laugh for the first time in five years instantly remembering the day by the river, when she had transformed a youthful Henry into a man.

Whilst Father Parsons voiced the blessing, a change began to take place in Henry - for the first time in his life references from the Bible held meaning, supplying the guidelines for a meaningful direction to his life, and crucially, a safe path upon which to tread it.

Henry's spirits rose to the point where he sealed the proceedings with a kiss, to the delight of all around, and none so more than the recipient - his new wife and true love.

With raised voice Father Parsons confirmed the marriage of Jane Taylor to Henry Winstanley, close friend of the King - and with that Jacob Wiley had heard all he needed to know…

Once the music had stopped and everyone moved outside, the merry-making was in full swing to everyone's delight – the girls now free to fight over Ben Drury again, the women remarking how wonderful it had all been with some even shedding a tear, whilst the men acted as a pack heading swiftly to the sanctuary of the 'Golden Goose Inne'.

Mrs Taylor, so overcome by the occasion, dabbed her eyes and declared loudly for all to hear, how she had always proclaimed Henry above Benjamin Drury, as the perfect match for her daughter.

Stepping out of the church, the new bride and groom took the opportunity to greet one another, and as with tradition, the lady was allowed to speak first:

"So, you still wished to marry me?"

"I always did."

"But, just look at me."

"I have never NOT been able to look at you."

"But, I have changed."

"I beg to disagree - your eyes are the same, as is the scent of your hair."

"But, I can no longer dance."

"So? – I have never been able to."

"Cast your eyes lower – my hip is broken."

"Then I shall fashion a brace from brass and springs – I have learnt much from my time in Europe. I have created something to impress His Majesty, and if you wait here a moment, Mrs Winstanley, I hope it shall also impress you."

"There is no need, Henry, you have impressed me enough already."

Then, as rain began to fall, and the crowd began to disperse, Henry performed a theatrical bow, and dived from view into the covered cart. For a while Jane was the only person watching, until what sounded like the gasping of an asthmatic beast lurking inside the covered cart drew the crowd back. Amongst the inquisitive faces was the olive-skinned head of the aforementioned Jew, bobbing to left and right, eyes straining for a clearer view whilst his mind busied itself on hatching a plan…

CHAPTER 23
LED BY GREED

Permit me to re-introduce the young and ambitious Mr Jacob Wiley, aged twenty-two, having fled his parent's nest of unquestioning Judaism in Cambridge with a plan to hitch-hike 54 miles due South for the unquestioning world of Finance in the City of London.

As you well know, Nature dictates that every one of us is the product of our parents, so it should come as no surprise that when a swindler and a drunkard conjoined 21 years ago there was little hope of a noble outcome. Jacob Wiley's first steps led him onto the path that suited him best – Finance – much to his impoverished parents' delight. Having flagged down a passing cart heading South the previous day, the penniless Jacob ensured his first steps onto that path by tipping the driver a silver coin before leaping on board. Yet, plans and reality often part company as did the cart and one of its wheels 14 miles later, tipping Wiley into a hedge bordering a world of unquestioning Feudalism just outside the market town of Saffron Walden.

We rejoin Jacob Wiley amongst the crowd in the rain, captivated by the sound of gasping emanating from the cart until it emitted its last gasp, and Henry popped out, pulling back the cover to reveal a pair of bellows and the pig's bladder it had been inflating. Buoyed on by the applause, Henry attached a chord to the bladder, and led it bouncing like a balloon up to the bemused Jane, before placing the chord in her hand and, in the manner of a magician, draped a silk sheet over the bladder then released it into the air. The crowd gasped, and some laughed, to see what was, in effect, a floating umbrella – the silk sheet protecting the bride from the rain.

"Congratulations, my husband, our King will be delighted you returned with such an innovative gift from Europe."

"Not from Europe, my dear, but an idea that came into my head in the sodden field of a German pig farmer during a particularly nasty downpour. The precipitation provided the inspiration and one of his pigs provided the bladder."

Ever the opportunist, Jacob Wiley knew how to take advantage of the slightest remark and on this day, he quickly put two and two together and made twenty-two, "Excuse me, Mr Winstanley, permit me to offer my congratulations."

Both Henry and Jane turned to acknowledge this stranger, curious as to his motives, "We thank you, sir."

At this point Jacob Wiley puffed his chest before declaring, "The King set me aboard His coach to introduce myself at your wedding."

Henry could not conceal his shock, "I am delighted yet surprised, sir; I had no idea my King held an interest in my marriage. Who can I say I am talking with?"

They were presented with a flamboyantly theatrical bow,

"Please permit me to introduce myself – Mr Jacob Wiley of Wiley Associates in the very heart of London. With your blessing I shall endeavour to ensure your idea is protected from any unscrupulous opportunist and believe me, there are many such individuals hereabouts. It is a sad, sorry scenario that happens with increasing frequency these days. Believe me, it is always prudent to protect your own interests. I know a few investors who would find your ideas eminently bankable…"

Henry shook Mr Wiley's hand gleefully, then clasped his wife's, "Did you hear that, my dearest, Mr Wiley here claims to have my best interests at heart."

Jane watched her husband's face, so on fire and without wishing to douse the flames attempted to add a modicum of caution,

"As do I, my dear."

"Yes, but Mr Wiley is promising great things."

"And I am promising wonderful things."

"Yes, but he understands me well, and will protect my future."

"Nobody understands you better than I. This day I made a vow before God to honour you for eternity. I never break a promise."

Jane watched Henry's turmoil, caught between the chance of Fame and the certainty of Love, and awaited his decision. When his eyes broke from hers and looked to the ground, Jane sadly knew he had made his choice. The slap of the gentleman's handshake sealed the deal - for Jacob was a man of ambition who had recognised a soulmate.

Jane watched Henry adoringly, seeing her husband so full of hope, and concluded that if it was due to Mr Wiley, then her instincts must be wrong, and this Mr Jacob Wiley must be a decent man after all.

Henry turned to address the crowd, "From this moment forth, I declare Henry Winstanley to be the happiest man in the whole of England!"

As the crowd broke into applause, Henry looked up and noticed the church clock was broken, "I have seen many spectacles of Nature these last few years from storms to St Elmo's Fire, but please tell me, Reverend, who in Heaven's name learnt how to make time stand still?"

The Vicar stepped forward, somewhat shyly, "Our clock stopped the moment you went away, Henry."

"Then just as I have returned, so shall your clock."

"If that be the case your name would be exalted by every soul of this parish, from beggar to nobleman, for each would have benefited from your greatness."

When surfing a wave of adoration, one can be prone to embellishment and Henry was no exception, delivering a bold declaration:

"To the good townsfolk of Saffron Walden – you will be the first to witness my work from which I trust every man, woman and child shall benefit. Today is the start of a grand plan I shall undertake to benefit your town, my country and my King. I declare that within one calendar month you shall once again hear the percussions of your clock and henceforth you shall hear much more from Mr Henry Winstanley Junior. I shall begin my work at first light. My apologies to your blackbirds however, who shall need to find a new spot to roost."

The crowd burst into applause again blocking Jacob's view, for being vertically challenged Mr Wiley took inspiration from the humble Tench – surviving as a bottom-feeder in the murky depths. For a while Jacob's eyes scanned wherever they could reach, and it was whilst performing that scan through the forest of legs that an alternative prey came into view, and being both crippled and solitary, Jane provided an easier catch, for after all - she was going nowhere fast.

Jacob approached nimbly - grateful the crowd's attention was focussed elsewhere.

"You are fortunate, Madam; your husband is a very clever man."

"I would rephrase that as 'very gifted'."

"Quite so, Madam, does he conjure such ideas often?"

"All the time."

Jane sensed Wiley's attitude turn on a pin, from charming to controlling,

"Then I must emphasise the fact, that your husband requires my protection."

"Do not concern yourself on that account, sir, I can protect him."

"I mean to say legally, in matters of business. Forgive me for asking, but what knowledge of business do you enjoy?"

"A good deal, sir, at my parents' tailor shop on the High Street."

"Modelling pretty dresses is a far cry from running a business, Madam."

"Indeed, sir - modelling requires a beauty beyond my present capability. I govern the accounts, balancing revenue against expenditure which have returned a healthy profit since I took control, plus I always pay my dues."

"That sounds excellent, Madam, for a small business concern in a small provincial town, but I deal with international businesses on a regular basis from my office in the City of London."

Jacob Wiley paused to perform a flamboyant bow (which if he knew this woman better, would be of no use at all), "Let me declare 'Wiley Associates in the heart of London' at your service, Madam."

Members of the crowd, intrigued by this flamboyant peacock, stepped closer. Irritated by the conversation steering in a direction she did not wish it to go, Jane intervened,

"Are you in pain, Mr Wiley?"

The crowd took a step even closer upon witnessing Jacob Wiley begin to shake, frustrated at being halted so soon into his pitch, "Excuse me, Madam?"

Before he realised what she was doing, Jane's hand was upon his cheek,

"Your scratch appears fresh and angry, poor man."

The crowd were now alongside, witnessing a man who courted control, being wrong-footed by an invalid woman.

"It is nothing, Madam…"

But Jane was not a woman to be distracted,

"A fall then…?"

Jacob gasped, feeling as helpless as a child, and worse still, being so closely observed,

"A fall? … No, no, I chased a thief and received this as a consequence – now, as I was saying, Madam…"

"Then you are a most courageous man, sir (and upon noticing his jacket) with a good taste in clothes. In fact, we sell an identical item at 'Taylor's The Tailor Shoppe' on the High Street…"

It was now Jacob's turn to divert attention, "Madam, forgive the intrusion, but it troubles me to think Henry's mind, although brilliantly inventive, may not support the business acumen required to provide you with a healthy financial base. The fame he will earn when news of his creations spreads across county borders, will attract a good many vagabonds of the lowest calibre. We need to discuss my position as business manager as soon as possible."

Despite being ruffled by his familiar manner, Jane had been gifted with a copious degree of diplomacy:

"Sir, as I have already indicated, you can be assured my husband shall receive all the business advice he needs, from his Manager."

Jacob reaction was instantaneous as one who had been wrong-footed, for his voice rose by a few octaves into the sort of squeak reminiscent of a farmyard pig:

"He already has a Manager? But I have not yet had the opportunity to introduce myself in that capacity!"

"Do not fear, Mr Wiley - you have the chance of introducing yourself to that Manager here and now."

"I have?"

"Greetings, Mr Jacob Wiley, my name is Mrs Jane Winstanley, Business Manager to my husband Mr Henry Winstanley."

The crowd laughed and Jacob, furious at being wrong-footed yet again, did his utmost to rectify the situation with another of his ingratiatingly deep bows,

"I would offer you my printed details but alas, my complete set has been spoken for, this day."

Jane stepped forward to deliver the coup-de-grace, which she accomplished with a degree of relish, stroking his scar,

"Do not concern yourself, Mr Wiley, I never forget a face."

The crowd moved even closer, intrigued as to the outcome of this altercation.

Jacob Wiley's pride stepped in to save the day projecting his voice for the benefit of the crowd,

"Thank you, Madam, from this day forth please refrain from merely mentioning Wiley Associates and instead call me Jacob. And please permit me to address you as Jane."

The crowd was now alongside, caught up in the drama. Making the most of the situation, Jacob Wiley performed yet another exaggerated bow and, to the crowd's delight, reached out to kiss her hand. But Jane swiftly took control, bringing a gasp as she thrust her hand into his for a gentleman's handshake. With a smile that conveyed her teeth more than any warmth, Jane concluded with:

"You may call me Mrs Winstanley, thank you, Mr Wiley."

The crowd began to murmur, comparing their thoughts, as the protagonists parted company, not as friends but as opponents who had each had a taste of the other's blood, Jane climbing into the cart alongside her husband whilst Wiley climbed into the bushes to spend time amongst the most non-judgemental creatures he could find.

As if to press home a point, as the kind-hearted owner of the cart arrived home to impress his wife with the tip he had earned from 'a charming, young man who needed a lift'. However being more worldly-wise than her husband, the wife tested the silver coin and felt it disintegrate into dust. Thus, the cynical wife remained a cynic, and her naïve husband grew a little wiser...

Life has a way of balancing a positive with a negative - the positive here being setting the ground-rules for the newly-weds' future together.

What should have been the most romantic day of their lives concluded as equally passionate but, in the wrong direction for as the cart departed, the fact it was covered allowed the opportunity to air opinions without fear of being overheard...

"What was I to believe with you being away so long? Your dalliances cut me to the core."

"What dalliances?!"

"Any normal man would have taken advantage of the freedoms on offer."

"I am not a normal man."

"Obviously. So, what is wrong with you?"

"Love - pure and simple. So, by your thinking, you believe love is wrong?"

"I do not believe *you*!"

"Or you *dare* not believe me."

"Do not put me on a pedestal I can never climb."

"Then fall down and join me."

"And live with an adulterer?"

"Be quiet, woman, how dare you judge me against the lowest forms of men."

"Such arrogance!"

"Then leave me to develop my 'ordinary' ideas on my own. The problem clearly is that I have experienced a new world, and you have remained stuck in provinciality."

"What does that mean – Provinciality - I have never heard it before?"

"I believe I overheard it recently, or I may have just invented it – I am good at inventing things."

"You are also good at deception."

"For all that is holy, woman, I expressed my feelings clearly enough in my letters!"

"What letters?!"

"The hundreds of letters I entrusted my father to give you."

"I have received no letters."

Henry pulled the cart to a stop, "So you remained loyal to me despite knowing nothing of my feelings?"

"Well at least, I now know a new word – provinciality."

The anger that was trapped under that cover suddenly broke free, leaving the young lovers to enjoy being in love once again.

"This cart will be too hard on your back."

"Then invent something."

Henry rose to the challenge, and by using his coat as a cushion and practising the ancient principle of 'Trial and Error', Henry finally discovered a position that turned Jane's cries of pain into groans of passion.

Many minutes later - allowing for Nature to run its course, they lay together content to hear the other's breaths and feel each other's moves. Jane was the first to speak,

"Optimism is one of your most endearing qualities Henry, but promising to design and build a clock tower within one calendar month is needlessly bold to the point of stupidity."

"Did you not hear what Father Parsons said? My name will be passed from lip to lip. My darling wife, from this moment on you must trust what I say, for I shall never promise what I cannot deliver."

Jane nodded, saying nothing, accepting having been gently put in her place. Expecting Henry to be angry, Jane was pleasantly surprised when he grinned conspiratorially pointing to his head,

"Remember when I said I have such wonders stored in here – well, the clock is already finished, here in my mind, and will soon start to be built - but that shall remain our secret. I shall prove it to you in good time."

"This is not how I imagined my wedding day – and no father of my groom present, although I thought I caught sight of him entering the church as we left."

"He did not know, I had no desire to tell him, and I felt no loss."

"I do not purport to know your father, but I do know you need to stop troubling yourself over him: you are your own man now, as much as you are mine."

"Dear Jane, you have wed a son spurned by a disinterested father – I should never wish that upon a child of my own."

The remark sparked a fear in Jane who replied nervously, "Henry, I have already told you that I can never bear us a child."

Henry aimed the cart for the long drive back to Audley End, "Do not fear, my love – I am child enough for both of us."

The rain became lighter, helping the mood as they entered Audley End ending their journey. Henry brought the cart to a halt alongside one of the barns, grinned at Jane, and raced inside greeting Jack who was passing on the way out calling,

"Congratulations, Mr and Mrs Winstanley."

Jack's youthfulness came as a welcome distraction before Henry rushed out of the barn, but to Jane's surprise he raced straight past her like an eager pup towards a tall, distinguished gentleman with the manner of one who had nothing to prove, overseeing building work at the side of the house. To her disappointment Henry ingratiated himself, proffering a deep and affecting bow.

Jack approached shaking his head, mixing words of wisdom amongst all his naïve clumsiness shouting, "Henry, what has become of you? Have you forgotten your beautiful wife already? You have only been married but five minutes."

Jane was grateful, sensing a useful ally in Henry's friend,

"Do not concern yourself, Jack, Henry is easily distracted, but pray, tell me who is that gentleman?"

"That is Sir Christopher Wren, Madam."

"Ah, then that explains everything. Henry has not changed, but I fear he is ingratiating himself when there is no need: Mr Wren may be gifted, but I believe Henry to be very, very gifted. Revealing too much of himself is lowering his status needlessly."

"Henry may be selfish at times, but he does have a good heart. And now he has a good woman to fill it."

"Jack, from this day forward would you please call me Jane? I believe I may have need of your support in the future. I realise that in many ways you know my husband far better than I."

At this juncture it is worth pointing out that ego in itself is not necessarily a bad thing - without it there would be no Kings nor Queens, performers would fail to perform, and songs would never be sung.

However, wars would still be fought and a person like Henry would be in danger of thinking too highly of himself...

Jane watched with some relief as Henry turned back to join her, leaving Mr Wren behind – yet, a frown had taken the place of his smile and his feet dragged across the lawn, having lost their former skip.

Luckily Jack came alongside to lighten the mood with a dose of youthful naivety,

"Come on, Henry, cheer up for heaven's sake. You have nothing to complain about. You are on good terms with the King, you have secured a most beautiful wife and above all else, you are exceptionally lucky to have me as your friend. Now put down your frown, and show us your latest wonder."

To their relief Henry managed a smile and uncovered the cart revealing two tall fishing rods on either side held in place by ornate metal harnesses supporting one lamp each – red on the left and green to the right. The applause brought a smile to Henry, but it was short-lived due to the grin on Christopher Wren's lips.

Jane intended no harm with her following words:

"Mr Wren finds your ideas amusing."

Henry's frown immediately returned, "One day he will regard me seriously, as will my father."

Jack retreated, sensing the newly-weds' need for privacy, however Jane refused to be downbeat,

"Smile, Henry, I want you to return the man who made me laugh but ten minutes ago."

Henry snapped out of his reverie, "And so I shall – climb aboard, I have created something especially for you…"

Jane squealed with delight grabbing her hat as Henry set Toby at a canter passing so close to Mr Wren he had to jump out of the way, and steered the cart to the water's edge, precisely to the spot where Jane had transformed him into a man five years and a lifetime ago.

Jane watched intrigued as Henry tugged at the cover, "Allow me to present my latest wonder…"

Removing the cover revealed another of Henry's contraptions – this time a drinks holder with a difference; for a bottle of lemonade was being kept chilled by a bowl of ice in its base,

"Let us toast an old friend – Lemonade toasted our courtship, so it is only appropriate that it should toast our union."

As they clinked glasses with eyes fixed upon one another, it was no surprise they didn't spot Mr Wren watching from a distance, his mind inhabiting a world beyond such trivialities before smiling, turning around and heading straight back for it.

Jane's observation had been accurate - Henry's father had indeed visited the church. But what she had failed to notice was Father Parsons handing over a purse full of money before cementing a deal with two bottles of sherry intended for Holy Communion.

Two glasses of sherry led to many more that night – for Henry's father had just fostered a deal ensuring his son would forsake his dreams of Fame & Fortune for a career of Piety & Poverty in the Church…

Upon leaving the church far tipsier than when they had entered, Henry's father and Father Parsons bumped into Mr and Mrs Taylor leaving 'The Golden Goose Inne' equally merry from celebrating their daughter's wedding.

And so it was that two men of the cloth joined a couple whose business profited from cloth and were invited indoors like long lost friends.

Meanwhile, upon heading outside to chastise the still-barking dog, Mrs Taylor noticed something she might otherwise have missed – tell-tale

stains of blood on her wall and a tiny strand of green cloth caught on the nail.

Staggering back inside, a sudden pain in her foot caused Mrs Taylor to fall, ruining her heel but exposing the culprit – the rosary bead now nestling in the palm of her hand...

For those of you concerned for Mr Wiley's predicament please fear not, for as in the old adage: 'The Devil Looks After His Own', the stage was set for the rest of our story – the rainclouds were shuffled off-stage granting Jacob Wiley a dry night's sleep before the Sun was wheeled in the following day to warm his cheeks...

CHAPTER 24

LOVE VERSUS AMBITION

Henry's workshop was his favourite place on Earth, where he could construct his constructions without fear of interruptions, and where we find him now, revealing his latest wonder to Jane - as soon to be revealed to the whole world (or at the very least, the folk of Saffron Walden)...

"That is wonderful! They move like clockwork," Jane gasped, watching the rotating planets in motion.

"Precisely, and this will form the basis of my new clock. All it needs is a place of shelter and the 'Winstanley Touch', to set it apart from the rest. Mind the spiders..."

Henry guided Jane through the curtains of webs that had accumulated over the last five years, to a table whereupon lay the plans of how his new clock design would look, including a huge glass lantern set on top.

Jane's eyes were as wide as plates, "It is wonderful – I have never seen such a thing."

"Nobody has - it shall define Mr Henry Winstanley as much as a signature defines its signatory. I expect crowds to be drawn to the church as irresistibly as moths to moonlight. Of course, my father will not approve."

"He is in no position to judge anything, look..."

Henry joined her at the window to see his drunken father drop from the carriage below, eject the contents of the previous night's sherry across the cobblestones, then stagger inside.

Jane laughed, "You see, my love, he is human after all."

"Aha, but a human possesses a soul, while he most certainly does not."

"Pay it no attention Henry - this day serves to lead to the next, when the plans you hinted at on our ride shall unfold."

"I must confess I have repeated them to myself so often - now the time is near, I am afraid the reality shall prove me lacking."

Jane sprang into action surprising him, taking hold of his hand and shaking hard,

"Do not think like that! I believe in the mind I fell in love with – do not dare let me down!"

Over the following weeks, Jane watched and heard her brooding husband toil over his plans, grumbling throughout until the grumbling turned to humming, the melancholy subsided and they became surrounded by excited gossip as they approached the town, picking out comments such as 'What in God's name is hiding under that cover?' 'Have they stolen our clock?' 'It would be foolish – the thing doesn't work!' The more he overheard, the more excited Henry became, knowing that he would be hailed a local hero, his name would be uttered on everyone's lips, and he would be held in the highest affection by all and sundry, reaching the ears of wealth and influence across the county.

Henry's baby remained hidden between the two lower spires, motionless like a corpse under its shawl awaiting the chance of Life. A huge glass lantern sat atop, protecting one hundred candles – apt for a place of worship providing a beacon of light for any poor souls seeking sanctuary...

The crowd fell silent as the sun eventually appeared from behind sleepy clouds touching them with its brilliance, instigating a great deal of gossip as to why young Henry had been commissioned to create something of such importance when his father had often been heard dismissing him as 'purely whimsical'.

Surely then, it was destined to be a disaster...

Moving through the crowd were two men sharing one aim – to profit from Henry's genius: - firstly the aforementioned Jacob Wiley on the lookout for any chance of advancement, and secondly a kindred spirit of sorts lurking beneath a beaver skin hat – Tony Tyler, the 29-year-old Reporter of London's first newspaper - The Gazette, eager to make his mark before his 30th birthday, now parting the crowd to reach the entrance. With his tiny eyes darting to left and right flanking a pointed twitching nose, Mr Tyler exhibited all the charm of a voracious shrew, and in the spirit of that creature required a full quota of news each day to satisfy his specific hunger – feeding a readership which was growing by the week. As one would expect of a shrew, his voice was clipped and highly pitched displaying a mean demeanour,

"Where is Henry Winstanley?"

Several fingers pointed up to the covered clock, so that is where he headed, pushing his way past several others without apology, leading the way for Jacob Wiley to follow in his wake.

At the top of the steps, they witnessed Henry's father adopting the role of Master of Ceremonies, bowing to the expectant crowd below and, with some degree of ceremony, tugging the cover away to reveal his son's magnificent, gleaming brass contraption. The crowd roared with delight, and Henry's father was equally delighted to receive such adulation.

Still a few miles away, Jane could hear the commotion, "What on earth can be happening ahead?"

Henry snapped the reins and Toby responded - grateful for an opportunity to run.

A huge cheer burst forth from the crowd as Henry's structure shuddered into life, surprising his father with its interplanetary waltz.

Henry's approach to Saffron Walden was akin to racing to a show after it had already started, as people raced ahead on either side of the cart, excited at the prospect of their new clock and exclaiming for all to hear:

"They say - it holds one hundred candles."

"Aye, unbelievable."

"And it shall play chimes on eight bells."

"I hear it has planets that shall rise and set with the Sun every single day."

"Aye, that Winstanley fellow is a genius."

"Quite so, although he is somewhat older than I expected…"

Henry's bewilderment increased as he finally thundered along the High Street dividing the milling crowd who were supplying comments that gave a clue as to what had already taken place, until the church finally came into view – the structure Henry had nurtured from the kernel of an idea to its present glory. His blood pumped harder as each new clue slotted into place until he was fit to burst, scattering the townsfolk until he skidded to a halt at the church gates at precisely the moment his gleaming galleon voyaged towards the giant brass Sun, where to the crowd's delight the cannon doors opened, and at precisely 13.00 of the clock it fired, causing an enormous strike of the gong almost hidden by the ecstatic roar of the crowd.

Father Parsons emerged from the church, greeting Henry with arms as wide as his smile, "Your father deserves much praise, Henry. It is a magnificent success."

"Where is my father?"

"He went home swiftly; I presume to tell your mother the good news."

He patted Henry on the back adding, "I have passed your fee to your father – £8 being a small amount for such an outstanding achievement, and by the way do not concern yourself on the other matter for I am confident you will be welcomed with open arms. You will be required to be ordained of course but have no fear, I am sure we can push that through swiftly - he was quite insistent on the matter."

"What do you mean by 'the other matter'?"

"After all your father has achieved today, I am sure there will be no obstacle to your enrolment."

"My enrolment into what…?"

Father Parsons continued unabated, lost in the joy of it all adding, "The man from the newspaper questioned everyone…"

"What is a news-paper?"

"I do not understand these things, but I am told they carry a record of important events of the day which are somehow written into a book of sorts for everyone to read."

Henry's mind was in turmoil, trying to make sense of this sudden madness, and realised a question he had to ask: "When you said, 'for everyone to read', does that include The King?"

Father Parsons faltered, fumbling for an answer to something he hadn't had the wherewithal to consider,

"I really have no idea, but I suppose if all of His Majesty's subjects can read it then the King would be entitled to read it as well…"

"Where is this paper-man?"

"He fled to catch the London coach."

To Jane's surprise, waiting in the cart, her husband suddenly ran up and grabbed her arm, "Wait here," then proceeded to sprint away down the High Street on a tide of fury, desperate to find a man he'd never met about something he was only just beginning to understand.

Henry had no idea what this paper man looked like, or indeed where he was, but the Vicar had mentioned the London coach and he suddenly caught sight of it facing him a few hundred yards down the road, as a man wearing a beaver-skinned hat was stepping onboard. The driver cracked his whip, the horses screamed, and with a flurry of hooves the coach sprang forward. Henry sprinted down the middle of the road trying to stop it with arms waving wildly, but the coach kept coming with the newspaperman on board, determined to leave town. Henry was fast but the coach was faster, and about to pass the church when to Henry's relief Toby appeared pulling the cart across its path with Jane at the helm forcing the coach to a shuddering halt in a cloud of dust. Tugging open its door Henry easily spotted his target – the beaver-hatted man looking smug cradling a bag pregnant with notes. Upon spotting Henry, he turned white and threw both hands into the air pleading poverty. Henry introduced himself and explained he was not a highwayman, and had no intention of stealing money which calmed the newspaper man somewhat until Henry demanded to speak with him, and the man threw up his arms once again,

"I have already spoken with Mr Henry Winstanley and have no wish to hear any nonsense from an imposter whose sole intention is to steal another man's glory. I have a duty to report the truth, sir, not the lies of a charlatan."

Being an honest soul, young Henry remained calm, not wishing to be heard tarnishing his father in front of a stranger.

Anything can usually be negotiated with the right incentive, so after promising an even bigger story over a free pint of ale, Tony Tyler's attitude changed, the colour returned to his cheeks, and he willingly followed Henry into the 'Golden Goose Inne'...

The landlord called Henry by name, thus providing Tony Tyler with a proof of identity, supported by the locals adding proof of achievement by hailing him a hero for returning their clock. Jane arrived shortly after and swiftly made a proposal of her own in her capacity as Business Manager - and so it was that over the course of two more ales, Newspaperman Tony Tyler learnt of Henry's exploits for the King of England. Layer by layer Tyler learnt about Henry's true character, and by the same token Henry learnt how a newspaperman's principles can be swayed by a free pint of ale plus the promise of an enticing story. Jane's principles however remained firm and she took her managerial role seriously, so it was in that vein when she announced:

"Mr Tyler, I suggest you follow my husband's future closely, and if you are willing to write supportive words and allow me, as his Business Manager the right of seeing them first, I shall not only allow you privileged access to Henry's future creations, but also his dealings with The King."

At the end of the meeting Tony Tyler asked each to supply a quote for his readers – a personal one from Jane (for he could foresee a market for such 'female-driven' articles) and a formal one from Henry to satisfy the thirst of his more traditional readership.

With quill poised Tony Tyler recorded their quotes, complying with the fashionable taste of the time by starting with the lady:

'I am proud to bear the name Mrs. Jane Winstanley,
offering my husband my hands to clean his clothes,
my arms to stir his food, my ears to hear his woes,
my eyes to witness his creations, my bosom upon which
to rest his head, and my wits to manage his business affairs.
I invite all to come and visit the fair town of Saffron Walden and
witness my husband's creations for yourselves. Mr Henry Winstanley
is in the ascendancy – catch his works whilst you can.'

In a manner of suitable import, Tony Tyler then turned to face Henry,

"Sir, are you able to add some comment as the auteur of this marvel?"

Henry thought on it for a moment – if his words were going to reach London, then it was best he choose them wisely:

'To the good townsfolk of Saffron Walden and the fair City
of London, I declare this important event marks the first of
many Wonders I shall devise for the good of my Country,
and the glory of my King.'

Then turning to his wife, he added privately,

"Hear me on this, my love - I know I am here for a reason, but I know it is not this."

Jane simply sighed whispering, "Be positive and accept praise whilst it is being offered Henry," and kissed his cheek leaving Tony Tyler with a degree of admiration not previously evident.

"Madam, I sincerely hope your husband appreciates the support you are providing this day."

"Sir, it is nothing more than the vow I have recently made."

"Then I hope his creations will deserve your confidence, Madam, and I wholeheartedly accept your proposal."

Thus, a perfect partnership had been formed where both parties provided for the other's needs – History proving such relationships to be the most enduring of all. Tony Tyler's quill danced like a fevered mayfly, as he swiftly composed a contract which was duly signed by all parties, and a copy placed in Henry's hand.

Despite Jane's protestations, Henry was in no mood for compromise and, fuelled by the ale, staggered to the church, commandeered one of its horses and rode like the wind to confront his father.

Henry barely remembered what happened next, his temper being so inflamed, but moments were to remain etched in his mind for the rest of his life – the banging open of the door to his home, his mother's face drained of all colour save grey, his voice growling 'Where is he, mother?', and her reply, 'Where he always is - in the Lord's House' - he remembered the screams of the young step brother he had never met, Miss Worthy past herself trying to calm him, and then being overwhelmed by fury banging through the back door, racing along the garden path, kicking open the church door and finding his father kneeling before Christ at the altar.

Henry placed Tyler's contract at his father's knees and waited - both Christ and his father remaining bowed in silence, "Read this, father – you should put faith in *your* son too."

Waiting for a response but receiving none, Henry strode away pulling the door closed behind him with a bang.

Henry never saw his father again. Having knelt in repentance all that evening and all through the night, until someone the following day realised, he was not going to repent any longer and removed him with knees still bent into a box they broke to accommodate his posture, and laid him to rest in the earth alongside his daughter, closer than they had ever been in life.

Fittingly, as the sun sailed towards the horizon that evening, all 100 candles atop Henry's clock tower were lit, thereby creating a beacon above the church lighting a pathway to God for any pious souls seeking sanctuary.

CHAPTER 25
TRANSFORMATION

Time passed, as it always does, healing scars and bringing about change, and so it was that Henry was finally freed from his father's anchor to explore his own interests in his favourite place on earth, which is where we find him today, in his workshop surrounded by his many creations in their various stages of development – being in this case a set of geographical playing cards, and a small brass fountain he was attempting to illuminate.

Jane burst in, breathless, bringing bustle to this place of calm.

"Stop, Henry, I have something to show you."

But Henry chose to remain as he was, lost in his world of small wonders,

"Look, Jane, what do you think of this? I learnt the technique at Versailles – to control water and light is the thing of Gods. It may be that I have finally discovered my true purpose on earth…"

Jane remained insistent – husband and wife occupying different worlds. "I have something to show you, Henry, please pay attention and read without prejudice."

The moment he saw the title – 'London Gazette' – Henry's interest was piqued, and he took hold, gingerly opening the pages as if they hid a coiled serpent.

Jane waited, afraid he might fly into a rage, so was relieved moments later when he laughed upon discovering the article was in praise of him and his church clock,

"A fine piece - Mr Tyler is exceptionally good at his job. We should allow him access to my future creations."

"I already have."

But once Henry turned to the following page, Jane's worst fears were confirmed seeing his face turn foul, having found a much larger article extolling the virtues of Sir Christopher Wren and his newly opened observatory at Greenwich.

Jane rushed to assuage him,

"Say nothing and do nothing - it is imperative that you embrace the fact Mr Wren is obviously a highly gifted and, more importantly, popular gentleman with both the public and The King. As your Manager I insist you embrace the populous and all its faults for, just like The King himself, we have need of their support."

Henry said nothing then turned his back on Jane to face the window, sitting perfectly still for such a long time she decided to leave him alone with his thoughts and headed for bed.

Henry stared at the stars, deep in thought, and as time passed and the stars slid across the night sky it was as if a moment of reckoning had been reached when he eventually rose to his feet, climbed the stairs and gave his sleeping wife a kiss before succumbing to a settled sleep alongside her, grateful for a partner who would take care of business matters, freeing his mind to create and dazzle once again, unhindered.

The stars eventually receded in favour of an excessively forceful sun the following morning, launching spears of light through their window, piercing the pillow upon which Henry and Jane were laying their heads, and returning him to the world of the present.

No words were needed as Henry headed for his workshop to face the new day and all its consequences. His first steps that morning passed under those who had become like old acquaintances now - the prominent figures of yore, set in oils, trapped in their frames lining the walls – Lords, Dukes, Landowners and Captains of War – passing underneath them without a shred of envy, for they represented positions in which he held no interest. However, upon entering into the Great Hall everything changed, rooting him to the spot and turning his blood cold. Barely 30 feet ahead, mounted above the fireplace for all to admire, was that magnificent portrait of Sir Christopher Wren, completely dominating the room.

As Henry stared in anger a brilliant shaft of light cast his reflection alongside Christopher Wren, presenting them both as if standing side by side.

He remained staring at it for a while, and the more he stared, the more Henry fancied his eyes were aligning with those of Wren, and he even dared to fancy they were now both becoming somehow equal.

Henry remained content until a familiar voice brought him back to reality with a thud,

"He looks grand, don't he? Your father thought so, God bless his soul. Now I must leave and prepare the breakfasts."

Jack had not intended any harm, but intentions and reality are two very different things. Henry raced out of the room only to return a short time later with a set of tools, erected an easel, canvas and stool in front of the painting then settled down to define himself, starting with his eye, then continuing to the brow above and the cheek below, all the while comparing his features to those of Wren.

As he painted uninterrupted in that huge space, Henry made sure that, on canvas at least, he was the equal of Wren, and as more brushstrokes appeared on the canvas so Henry's confidence grew, and as his confidence grew Henry spoke to his portrait, in whispers at first -

"Greetings, Henry…

Greetings, Henry Winstanley…

Greetings, Henry…Winstanley …Junior"

then growing louder and bolder,

"Greetings, Mr Henry Winstanley Junior…"

until his image was complete,

"Greetings, Mr Henry Winstanley…. GENT."

Thus, Henry had declared himself a Gentleman – thereby setting himself a code of conduct to live by…

CHAPTER 26
ANCHORS AWEIGH!

As the parents amongst you will know, any child of imagination suddenly relieved of parental control requires a firm hand on the tiller to remain on course, for Fame may well bring Fortune, but it comes at a price – the sizeable amount of unwanted attention threatening to distract...

Jane was the rock Henry needed - a footing on which to stand firm against a swell of events that could engulf them both.

It took time before Henry was to fully appreciate what, for now at least, his wife was offering without question...

Time had passed – only a couple of years faithfully acknowledged by Henry's planetary clock striking each hour of every day in a fashion special enough to have attracted a regular flow of visitors to witness the hammers playing hourly chimes on eight bells, whilst the Moon and Sun rose and set each day on 'Mr Henry Winstanley's First Wonder' – as advertised in 'The London Gazette' – bringing a steady income to not only swell Father Parsons' donor-box, but fill 'Taylor the Tailors' tills and drain all the barrels of ale at 'The Golden Goose Inne'.

Each visitor to the church had to first pass Jane seated at the door and drop coins into her box, before going on to pass Henry staring out from his self-portrait on their way to the stairs leading up to the observation platform for his amazing planetary clock. Thus, every single visitor saw what Henry looked like and gleaned a taste of his talent to tell others, thereby quickly spreading his reputation by word of mouth across county divides.

After enjoying the clock, the visitors descended a different set of steps leading to an array of Henry's other items for sale, including metal engravings of Audley End in various sizes priced accordingly, plus packs of his hand-drawn playing cards displaying exotic images of distant lands he had never visited, supported by informative rhetoric he had composed from books, gossip and thin air.

Unknown to them both, as the visitors filed out of the church, they were intercepted around the corner by the irrepressible Mr Jacob Wiley offering to insure their purchases in return for cash. Wiley managed to secure a decent trade that afternoon, but the main prize was still deluding him until, to his delight, Henry stepped outside followed by a group of female admirers fussing about him like a flock of hens vying to devour any crumbs left over from his success. His laughter permeated through to where Jane sat, managing the takings, aware of the capricious nature of the giggling girls surrounding her husband but putting trust in his behaviour as a gentleman – as declared on all of his recent works, now and for evermore bearing the signature: 'Henry Winstanley Gent'.

Jacob Wiley felt the opportunity had finally come to take his chance, so whilst Jane's attention was elsewhere, he slid away from signing contracts for one form of Insurance to benefit from a far more important one – namely his own and slid alongside Henry.

"Greetings, my dear Mr Winstanley, I wonder if you have had time enough to consider my proposal since our last meeting?"

"Sir, I must confess to my wife's caution on the matter."

"I sincerely agree with her. 'Caution' is the word I would always recommend for matters of finance, so in that respect your wife's attitude reflects my own. Please may we talk?"

As Jane was surrounded by customers buying their goods, when she spotted Henry passing the church with Jacob Wiley in tow heading towards town she could not follow, being forced to put her complete trust in her new husband.

Henry led Wiley to the 'Golden Goose Inne', as good a place as any to conduct private conversations – the locals creating so much noise, nobody could possibly follow what anyone was saying beyond their own table,

"Mr Winstanley, your genius was on display today for all to see in this most modest of market towns, but in the wider world markets dance to a different tune – diamonds glitter brighter than stones so I predict your genius will accumulate greater wealth in a bank than a barn, so I hereby offer my financial acumen accumulated from my time spent in London's banking Centre, to ensure your wealth is protected from variations in the market and other unforeseen events."

Henry was astute enough to recognise an unremarkable offer in the making, until something Jacob Wiley had said sparked his interest,

"What do you mean by 'other unforeseen events'?"

Jacob shuffled his seat closer - Henry had taken a tentative bite of his bait and he was now preparing to strike,

"Presently, you appear to be in robust health but consider if you should suddenly fall gravely ill or, God forbid, die: my contract will guarantee your wife an income from all your hard work to support her for the rest of her life. Tell me, how could you not consider that a worthwhile investment?"

Henry remained silent throughout concentrating solely on sipping his beer, causing the increasingly agitated Jacob Wiley to fiddle feverishly with his rosary beads, believing at any moment he may lose the greatest opportunity of his life.

After nine more minutes of impasse, Wiley spoke in desperation,

"Forgive my indiscretion, but would you prefer a different drink? A spirit? Or mead, or in fact - anything?"

"Do you think a fountain on this table would be popular?"

Wiley was completely thrown, not understanding a creative mind, so he spoke out of a need to say something, anything, no matter what,

"Excuse me?"

Henry remained oblivious of Wiley's dilemma, speaking from his genuine concern for an answer to his own, "I am considering a fountain with coloured water illuminated by candlelight?"

"If you believe in it, Mr Winstanley, then I would wage money on it being so."

"So be it - being a man who values money above all else, by saying you would risk your wealth on an illuminated fountain means you have made my decision all the easier…"

"Then you shall agree to my proposal…"

"Then I shall produce illuminated fountains!"

Wiley gasped in exasperation -

"And, if you can guarantee the protection of my wife's income, I will agree to your contract."

Jacob was beside himself with joy, beckoning the Landlord for two large brandies in celebration of guaranteed wealth.

However, as we all know, wealth is never guaranteed, and neither is the money that makes up that wealth, for whilst counting his takings at the end of the day, the Landlord was surprised to feel Jacob Wiley's coins disintegrating between his fingers.

Meanwhile back in the church at the end of a fruitful day, Jane was struggling to rise from her seat - so distracting was her pain that she failed to notice a wealthy Lady of 45, still handsome enough to pose a threat, regarding her from the doorway only 15 feet away, as a cat watches a limping bird.

Ignoring the young clergyman who was trying to close the door, this Lady calmly observed all the merchandise, including Henry's self-portrait, before gliding over to address Jane,

"Tell me – is that portrait a good likeness of him?"

Jane tried to hide her irritation at such a late customer,

"It most certainly is, he completed it barely four weeks past."

"Good, then is this how he looks today?"

"Aye, My Lady, it is."

"Interesting. I thought he would have been somewhat older considering…."

"No, I can assure you this is how he looks presently."

"And how can you be so sure?"

"Because I see him yawn each morning, urinate each day whilst scratching his head, and snore himself to sleep each night."

"Aha, I see you are attempting to make him less appealing, but I can assure you my intentions are strictly honourable."

"I never thought otherwise. Being his wife, I can assure you he is not worth the chase."

"But you obviously thought he was at one time - maybe when you were up to the challenge… My name is Lady Birch and I wish to examine your husband's handiwork."

"By all means, Lady Birch, take as long as you wish."

To Jane's irritation, Lady Birch took her at her word, and knowing Jane was in pain made a point of slowly examining each piece of merchandise until arriving at the playing cards, which she turned in her hand, intrigued by the sketches of exotic figures inhabiting exotic lands.

Jane stepped in to explain: "…Playing Cards, Lady Birch. They depict many exotic lands and their peoples. We are finding them extremely popular amongst our more well-heeled customers –those who may not have travelled to such lands, but would prefer others to believe that they had."

Lady Birch remained stock still, considering Jane's advice for a moment, then spoke as if instructing a maid, "I shall take three."

And added as Jane shuffled towards the table to wrap them -

"My coach shall collect them tomorrow mid-day…along with Mr. Winstanley."

The fact Father Parsons arrived straight after wearing a beaming smile, stopped Jane from reacting aggressively, and so Lady Birch climbed on board her coach none the wiser regarding the upset she had caused, and quite frankly, would have been concerned not a hoot anyway.

Inhabiting a world of naive purity Father Parsons was oblivious to Jane's irritation, reaching out his hand with a smile and remaining in that regard long enough for Jane to understand his intent, handing over a pouch containing his percentage.

No sooner had the Vicar scurried away with his takings, than Henry arrived smiling with Jacob Wiley at his side,

"My dear, allow me to re-introduce this fine fellow."

Jane was taken aback, and more so as Henry embraced his companion,

"Mr Wiley tells me his speciality is Assurance, and he assures me that he can protect us both from 'unforeseen events'."

Jane tried to compose herself under the gaze of Wiley's stare,

"I see. Now, I am tired so (and she followed by pointing her stick at one, then the other) you can go, and you can help me clear all of this."

Jacob bade his leave, fascinated by and attracted to this strong woman, and returned to the gap in the hedge that had served as his bedchamber for the last two nights, and was about to serve him in that capacity once again.

Aboard Henry's cart enroute to Audley End that evening, Henry took advantage of his wife's exhaustion, as she succumbed to the bouncing cart,

"I have an idea to place small, illuminated fountains upon dinner tables. Inns will love them and customers will obtain a thirst from the sound of their tinkling. I told Jacob and the landlord of The Golden Goose - they both thought it a very good idea, as I hope will you."

But Jane was on the brink of sleep,

"Do not confide in him any more...."

Henry wasn't sure which 'him' Jane was referring to, and was too tipsy to worry anyway, but added for good measure:

"I should like them placed upon every dinner table to impress our guests."

"But we only have one dinner table, Henry…"

"At Audley End yes, but trust me, we shall have room for banquets at our new house."

Jane didn't respond as she slid into sleep much to Henry's relief, for his mind was like a racehorse sprinting to the finish line - and ultimately the answer to his purpose in life…

Things began to move at a pace for Henry thereafter - his engravings and his playing cards proving so popular, he could afford a house nearby set in its own grounds and built to his requirements – requirements of a most bizarre nature, unlike any other in the Country, and indeed the World, with the aim to boost not only his wealth, but his Fame, and more importantly for Henry - his place in History.

CHAPTER 27
ACCELERATION

F*ame and Fortune are fickle bedfellows, never to be trusted - chase them and they will disappear in an instant, leaving you alone with your ego and nothing to feed it on.*

This day we find Henry and Jane enjoying a rare moment of reflection, on the banks of the river whilst fishing...

"A spinster came by yesterday, lured by curiosity of your fame and of you. Then like that fish she circled a short while until she could resist no longer, took the bait and now we must run with it…"

"Indeed?"

"She wishes to hook you, tomorrow."

"Do not fear, my love, I can be slippery as a fish if needs be."

"What concerns me is if your 'needs' change sides."

"You speak as if she is the trout, and I am but the poor fly."

"Just be cautious, my love, and play the long game…"

The following mid-day, the Lady in question was watching from her Manor House, licking her lips at the sight of her 'catch' approaching, secured in the trap of the handsome coach.

Springing to her feet she swiftly dismissed the maid before checking her appearance in the mirror.

Inside the coach the naïve Henry remembered his wife's wise words–

'*… use your cards wisely, my love, make her believe she is playing you, but beware of her tricks for she will be skilled in many more than you…*'

Henry stepped from the coach making his way to the door, sensing throughout that he was being watched, or more accurately that he was being stalked. Entering the huge empty house Henry understood how a fly must feel stepping onto a web.

Henry had never met a woman like Lady Birch - it was difficult to tell her age for she did not walk so much as glide across the room in a deliberate attempt to dazzle him, in which she was beginning to succeed - Henry being drawn to her slim arms and elegant neck sporting a carefully placed beauty spot for him to remove. Henry felt quite giddy as his eyes followed wherever she led until drifting behind the curtains she became invisible, providing an even more alluring catch.

And so, in his pursuit of glory, Henry was allowing his morals to take a back seat for a while.

Meanwhile, despite hiding behind the curtains, Lady Birch was well aware of Henry's complete attention and waited just long enough to maintain that attention before skimming back across the room like a dragonfly, falling eloquently into his arms, and planting a kiss upon his lips. To her disappointment, Henry did not respond, finding the smell of her perfume too much to bear and moreover, with an absence of emotion felt as if he was pressing his lips against one of King Louis' automatons.

Being unused to rejection, Lady Birch turned increasingly predatory enticing Henry into her massive drawing room, and as she danced and twirled, Henry put all his suspicions aside and politely sipped his tea before falling into a form of trance, suddenly unable to see Lady Birch clearly, but still aware of her figure weaving across the other side of the room performing movements of the hips the likes of which she knew Jane could no longer compete. And as she glided and span, her eyes never left his - her neck as a gyroscope keeping her sight fixed upon him at all times, no matter which direction her body took.

"I must admit to a soup-con of jealousy, sir."

"Really - how so?"

"You and your wife must dance every night."

"We do not dance at all, my Lady."

"My goodness, what a waste," she replied, feigning surprise.

The fact he did not rise to her bait was a credit to Henry for who knows what indiscretions he would have felt obliged to perform. In fact, Lady Birch had wasted her trump card in overplaying her hand – the sight of her hips had reminded him of his wife, and as such all the guilt that came with it.

Aware of drifting under her spell, Henry emptied his mouth of tea and concentrated instead on remembering his wife's words of wisdom:

'Keep your eyes on her cards and your hands on your own, and she will soon be eating out of yours. Show her your engraving of Audley End. Ask to see her Manor House, then offer to preserve it on copper plate, and most importantly, do not forget to mention the envy it will cause to her guests. Charm her, my love, and she will never realise how much work she has saved us, and how much wealth she has granted us.'

Henry duly asked as Jane had instructed, and found her idea worked - Lady Birch was delighted, and Henry accepted the offer of immortalising her house on copper plate.

Thus, the meeting had ended to Henry's satisfaction - envy having sealed the deal.

Finally, as he bid Lady Birch farewell with a final kiss on the hand, he received a handsome payment in his and as she leaned into his face, the offer of so much more…

But as her mouth drew near, Henry remembered the closing words that had come from his wife's: 'Finally, my dearest love, remember how much I both love…and *trust* you.'

The fact he did not succumb was a credit to Henry, for who knows what indiscretions she had in store for him.

However, the sight of her hips had reintroduced Jane into Henry's mind and as such, the love with which Lady Birch could not compete.

In declining Lady Birch's advances, it seemed to have the unintended effect of increasing her ardour once more, taking all of Henry's considerable charm to avoid her embrace as he retreated from the room, and swiftly headed for the front door when, to his great relief, two friendly faces appeared – Jane, looking vulnerable and Toby looking too hungry to care.

Keen to escape, Henry leaped onto the cart and was whisked away in far less opulence than how he had arrived.

And as he gazed at the woman beside him, her broken hip swaying from the movement of the cart still managing to distract him, Henry felt as if he were the luckiest man in the world.

Yet, once Audley End appeared on the horizon, Henry's mind began to wander – he was indeed relieved to have escaped Lady Birch's clutches… but what would have happened if he had not…would he have succumbed to her allure?

For the moment at least, Henry was relieved he had not…

CHAPTER 28

THE HOUSE OF WONDERS

H*aving gained such praise for his engravings and playing cards, Henry now felt confident enough to expand his horizons...*
Provincial enterprises held no interest for him anymore, but London did – Fame and Fortune were still his goal and if he was going to find them anywhere, then it was sure to be in England's Capital City...

On Piccadilly Circus, in the heart of London's Theatre Land, children could be seen on street corners holding up 'The London Gazette' carrying its clear headline: 'HENRY WINSTANLEY'S REMARKABLE HOUSE OF WONDERS'.

A smart sedan chair drew up next to one of the boys and a decorative glove snaked out presenting cash in return for a copy of The Gazette.

Inside the chair two secretive gentlemen regarded the headlines, their heads hidden under wide-brimmed hats,

"His popularity is in the ascendancy," said one.

"A wise man would ally himself with such a man, especially if his own popularity be on the decline," remarked the other.

The first man nodded in agreement, and the jewel-encrusted glove we have encountered before, clasped a staff and tapped the roof – alerting the chairman to set off at a trot in his eagerness to maintain privacy for his two very important passengers.

Within the hour that same glove had been discarded, and its owner was taking snuff in the House of Lords watching his Chancellor, Sir John Ernle, finishing a bloated speech concerning budgetary reforms:

"… therefore, Your Honours, while I fully understand His Majesty's wishes to the contrary, I propose that we reject any inclinations to fund another war with the Dutch…of which we do not have the financial means to support anyway."

As the echoes of 'Hear, hear' and 'Ayes' resonated around the House, Sir John Ernle's likeness was being sketched as an exaggerated cartoon by the frustrated King. As was custom, the Speaker of the House took control,

"The 'Ayes' have it. This Parliament will no longer fund a war with the Dutch. I therefore propose we begin negotiations with the Dutch Government towards a peaceful settlement as soon as is humanly possible."

King Charles swiftly drew a line through the cartoon's throat as a chorus of 'Ayes' resonated around the House. However, everything stopped the moment The King rose to speak,

"Gentlemen, I find myself once more looking across a sea of familiar nodding heads that, had it not been for my pardon, would have been nodding in a basket like my father's at the execution brought about by your decisions. Yet where common decency would demand humility, you continually strive to not support me. Therefore, it leaves only one option at my disposal…"

As soon as The King clicked his fingers, a mighty clicking of boots echoed around the House as uniformed sentries jumped to attention, sending a wave of shock rippling across the jowls of the fleshy MPs.

After a moment's pause, The King continued,

"My one remaining option is to invite you all to a banquet I shall host at Audley End the morrow. And before you leave, I would like to add that, although there is absolutely no obligation on your part to go, I shall make note of all who fail to attend and have the names circulated amongst my Generals."

The bang of his chair resonated around the chamber leaving His Parliamentarians either bemused or fearful – depending upon each one's role in the severing of his father's head.

CHAPTER 29
PREPARATIONS

We find Henry and Jane setting the final pieces in place for the most fantastical house in England, leading a team of workers readying the unusual devices for the Royal Opening: metal rails supported twenty feet off the ground on a wooden bridge being secured in place, a huge organ being tuned, inspecting the giant windmill for channelling river water into the garden's artificial stream, and contraptions being tested including Henry's favourite - tabletop illuminated fountains…

"My father would have been proud to know this day his son had knelt in prayer to God – but it was only to deliver the Sun for the morrow's opening. He would have hated this House of Whimsy."

"Well, I love my new house – hardly a normal birthday present, but I've grown used to nothing about you being normal!"

"I hope it also proves useful for the King to impress his detractors."
"Most important of all will be how Mr. Tyler chooses to print it - times are a-changing and so must we - one or two poorly chosen words will overturn all the fine work you have achieved here."

Meanwhile, just over one mile away at Audley End, King Charles was playing the genial host, supplying increasing amounts of the richest food and finest alcohol to his plump Parliamentarians. Relieved at His Majesty's apparent geniality, they capitulated to His wishes, not daring to refuse any of the excesses placed in front of them. The King smiled throughout, but behind that smile was a bitter man intent on revenge, remaining as sober as his servants throughout the proceedings, whilst his guests grew drunk on vintage wines. King Charles took great pleasure in dangling a particularly rare chunk of steak against a pale-skinned Duke's plump cheek, allowing drops of its blood to run symbolically down his pale and chubby neck,

"I am so sorry, my dear Mr Harrison…," said the King, making a show of removing Thomas Harrison's napkin to wipe the blood clean, "…However it does remind me of how we first met regarding my father, of whom I believe you were not enamoured. Does it not remind you also?"

Thomas Harrison juddered at the memory, upsetting another drop of blood to run down his reddening neck.

"My father was confined to History in a most bloody manner, yet History shall also remember His Son for granting pardons to many who had brought about that atrocity. Some will feel the new King too benevolent in that respect, yet others will deduce He did it for the good of His Country and kept any revengeful thoughts to himself, for that is the correct response of a civilised society…," then turning to two MPs sitting alongside, "… is that not true, my good Lords?"

One of them juddered, "Your Majesty, it is only fair to consider my position. I was only carrying out the orders of the highest authority in the land."

"Really?"

His colleague immediately jumped to clarify his position as well,

"Me too, Your Majesty."

"The highest authority you say? Well, as we all know, the highest authority is God and what you both committed was High Treason, which is punishable by death."

Both MPs shook as one,

"Your generosity is beyond doubt, Sire. As is your mercy."

The King outwardly smiled and, to their horror, held up two huge pork pies,

"My apologies, Your Highness, but I regret I cannot manage to eat any more."

"Is your neck too tight?" asked the King, playfully.

"Please accept my apology too, Sire."

"I am certain you would not want to be so rude as to decline such a generous offer, especially as most of your tenants dine on mere bread and pottage. So, eat up all of you, I have brandy and pies to follow."

As the bloated guests moaned and groaned, the King beckoned Captain Smith over, "See to it that those two arrive in luxury by coach but depart by cart."

"I understand, Your Highness."

"Good man and have them taken straight to the Tower, where they shall remain at this King's displeasure."

Thus, a Royal command was sent with a wink and received with a salute.

CHAPTER 30
THE GIFT OF INDEPENDENCE

Jane gasped – seeing only a tinge of red through her velvet blindfold whilst hearing someone approaching, she believed to be a man due to the heavy feet and deep breathing:

"Happy Birthday, my love…"

The stranger took her by surprise, removing her walking sticks, thereby compelling her to topple into his arms which guided her onto a cushion,

"And because you dislike being dependent…"

Jane's blindfold slid away to reveal her birthday present, bringing a burst of giggles – a motorised wheelchair powered by clockwork with her husband's whimsical accoutrements attached – items he had designed, plus others of King Louis' toys: an iced bottle holder for her lemonade, a hook for her spectacles, a cash till plus abacus for her accounting, plus the necessary quills and ink holder for recording those accounts, and to top it all the balloon umbrella, tethered to the handles keeping her dry in style.

A small group of well-wishers was arriving, distinguishable by their dress – from the 'Sunday Best' efforts of local farmers and villagers, to the fashionable frocks of city dwellers from as far afield as Liverpool, attracted partly by word of mouth and bolstered by the articles of Mr Tony Tyler spreading word from across the streets of London.

A smattering of applause mixed with polite laughter broke out, as Henry invited the crowd to watch him demonstrate the workings of the chair – the pumping of the armrest setting the clockwork motor into motion, propelling his wife as if by magic without need of either servants or horse, and all the while providing humour for the folk to spread across their neighbourhoods, increasing Henry's fame tenfold.

And as sure as wasps are attracted to jam, Jacob Wiley hovered in the background waiting for an opportunity to present Jane with a personal birthday gift, however inappropriate, knowing he was safe to do so in plain sight of so many well-wishers. Jane was taken off guard and, not wishing to cause a scene in front of so many people, opened the wrapping to discover a beautiful, and obviously expensive, hand mirror.

The sun reflected straight into her face, providing Jane with the excuse to wince,

"I would not have expected such a personal gift, Mr Wiley."

"I hope in time it shall serve to help you think better of me."

"At least I shall be able to better see what is happening behind my back. How appropriate, Mr. Wiley, thank you."

Jacob Wiley retreated blushing brightly at her very personal jibe, as the crowd around him laughed, without really understanding why.

A fine coach arrived depositing the first of the titled guests - the Earl of Suffolk, who approached Henry proffering his hand in welcome,

"My dear Henry, who would have thought it – welcoming me to your home having worked so long for me in mine."

Henry knew to be gracious, for after all entitlement comes with a title,

"It was always my intention that it would be thus."

Having expected a warmer welcome, the Earl bowed briefly, having been put in his place in front of the swelling crowd. Jane's keenest welcome was reserved for the man who had helped advertise this day,

"I have been true to my word, Mr Tyler, The King will arrive shortly, as I promised."

"Absolutely, Mrs Winstanley - His Majesty was most insistent upon my presence to record everything – 'Warts and All' – as his nemesis Cromwell once said."

Henry stepped forward, extending his hand, "I trust you will allow this day a fair degree of prominence in your paper?"

Tyler shook his hand with enthusiasm, "You can be assured of that, Henry – I have discovered that my readers show great interest in anything with a Royal connection…as obviously have others…"

Tyler had spotted a handsome, finely dressed gentleman alighting from a coach also carrying quill and paper, intriguing Henry,

"Another Man of Words?"

"Indeed, but I wager the words of Mr Samuel Pepys shall outlive mine."

Henry took note, observing Mr Pepys moving towards the entrance, his eyes moving to left and right – not in a threatening way but as an interested observer, already noting observations with pen upon paper, between interruptions from eager dignitaries obviously keen to be included in his diary.

Henry knew not to overplay his hand, remaining in the background in the hope of being presented with the chance of being included in Samuel Pepys' celebrated diary too.

As he watched Mr Pepys chatting amiably with guests of importance, Henry could not help but wonder if this was what Fate had in store for him - the noble thing that would seal his Fame for eternity…

"His Majesty comes!"

The servant's shout snapped Henry out of his reverie, and created a kerfuffle amongst all the guests, be they dignitaries or farmhands, there to enjoy the chance to rest their limbs awhile. Tony Tyler moved faster than Henry had ever seen, gathering papers, quills and ink before harassing his way to the front of the queue as his profession demanded, breathing deeply as he composed his thoughts trying to remember the most important questions he needed to ask of the most important person in the land.

It dawned upon Henry that this whole occasion, which he had worked so hard to achieve, was passing out of his control and into the hands of Fate plus the King, and all his cronies, each of whom had their own needs from the day – with some happy to settle for nothing more than keeping their necks intact. As he looked across the excitable crowd Henry's eyes settled upon the one person who remained serene throughout, as regal as any queen, seated in Henry's splendid chair - a birthday gift unlike any other.

Rushing to his wife's side, Henry gripped her hand more nervously than he had expected, feeding off her strength as Royal horns blasted, the crowd burst into cheers and their King finally arrived in all His glory flanked by uniformed outriders making the ground tremble as they thundered to a halt.

The King stepped from His coach into a sea of swirling subjects made euphoric by His arrival, rushing forward as waves rush up a shore to greet the rocks.

Surfing this wave of hysteria Henry thrust his hand into the air commanding the House of Wonders into life, firing up a thunderous organ bringing scores of candles to life within its roof lantern.

Jane swiftly pumped the handle on her chair three times, releasing the clockwork spring propelling her as if by magic towards the King, passing Tony Tyler and Samuel Pepys on her way, both wriggling their quills bringing their observations to life - her chair coming to an elegant halt at the King's feet who graciously kissed her hand,

"Am I correct in understanding that birthday congratulations are in order, Madam?"

"Indeed, Your Majesty, this is one of my presents and we invite you to enter my other gift as our first guest."

King Charles was gracious in His acceptance and, with a mischievous grin beckoned his guards to open all the coaches allowing the bloated Parliamentarians to spill onto the field and stagger towards the House.

After checking Tyler was recording everything, Henry approached the King and presented Him with a ceremonial sword,

"Fear not, this is a harmless sword made for the theatre, in keeping with the discoveries you are about to encounter here," adding upon a smile from the King,

"Welcome to my House of Wonders, Your Majesty, inspired by yourself via the European journey you set me upon, and your wish that upon my return I should benefit England with their wonders. I sincerely hope that you find my time was well spent."

The King leaned in closer to whisper: "If it delivers on that promise, it shall serve my purpose today admirably." Then, addressing the crowd, "To all who are gathered here - I am proud to introduce this House of Wonders as a fine example of two traits that benefit my country greatly – Invention and Ingenuity, both of which Mr Henry Winstanley Gent here possesses in abundance."

With all due ceremony, the King sliced through the giant banner suspended across the entrance sending ripples of applause throughout the crowd, whilst Jane commandeered her wheelchair to a turnstile beside the gate, for the selling of entrance tickets.

Henry rushed over to Tony Tyler for a private word, "Mr Tyler, please make note that The King gave thanks to 'Mr. Winstanley *Gent.*' for henceforth, this is how I wish to be known."

Meanwhile the King approached Jane acknowledging the admission charge,

"My Lady, I trust that, as my head is already placed upon the coins, I shall not be expected to also place one into your hands? I have had to learn the art of thrift as my Heads of Government disapprove of paying for my pleasures." And upon Jane's smile adding, "And here they come - every one of them stuffed like a pig and not a smile to be had between them - permit me to introduce my Heads of Government…" adding as the MPs staggered through the gate with much swaying and belching, "Is it not comforting to know the future of England is in their hands?"

King Charles then turned to face the crowd and extracted a one shilling coin from his purse holding it high for all to see, pointing to his image on the front:

"If each and every one of you were to extract a one shilling piece and hold it close, you shall witness my face is sporting a smile, which I trust shall be shared by all who enter this house with a willing heart and an open mind."

Both Samuel Pepys and Tony Tyler immediately set their quills on a merry dance recording his every word for their readers.

A Great Lantern designed to attract travellers like fireflies was proving its worth as the line of bloated MPs passed beneath, to the march of a church organ powered, like everything else here, by a giant windmill. As each wealthy Parliamentarian reluctantly dropped one of their coins into Jane's tin, Henry couldn't resist a smile at the thought of his father, for this house with all its frivolities would have infuriated him, its popularity would have dismayed him, and the addition of a church organ would have been the final insult.

"Do you see the short, plump one?" asked the King leaning to address Jane, "I have the measure of that one – despite his constituency suffering dreadful harvests of late, his belly somehow manages to expand ever greater."

"I always find the heavier they come, the harder they fall, Your Majesty."

Jane and The King shared a private laugh, which did not go unnoticed by her husband who swiftly leapt forward on a wave of jealousy to join them,

"My first trick is designed for one as plump as he. Come and enjoy his downfall, Your Majesty!"

Henry took control, leading the King away, leaving Jane to cope with the sudden rush for tickets alone, as a few feet away the speculative Jacob Wiley made the most of the situation, making his way along the queue offering what he would never be willing to part with – all details on his insurance policies, in return for cash.

Upon gingerly entering the house, the MPs were unnerved by the sound of children giggling in the shadows, and jumped three seconds later, as a piercing scream came from behind the curtains next to them. Up above, on a level reserved for the most special of guests, The King and Henry observed the happenings down below in complete secrecy: the plump MP shuffling along the hall, (for his pronounced waistline allowed no more than a shuffle) before noticing a plate of food floating in mid-air, a couple of feet above his head. Having no shame, the MP reached for the food, a trapdoor opened beneath, and with a shriek he dropped through the floor.

Henry trembled, concerned he had gone too far, until the King tapped him on the shoulder laughing, "Splendid, My dear Henry!"

Below them another MP stooped to pick up a coin eliciting a laugh from the King, "My Treasurer can never resist money."

"If my trick succeeds, then it will prove to be his downfall, Sire."

And, so it proved – for upon reaching the coin, the carpet flew from under his feet sending the Treasurer to his knees, amusing The King greatly,

"Thank you, Henry – the last time I saw him kneel was with my blade upon his shoulder. I am still not sure he deserves to be a Knight."

"Your Majesty, let me lead you up the garden path, as it were."

"I hope you have more tricks up your sleeve, Mr Winstanley, for I shall never tire of humiliating those who betrayed my father."

"Then look to your right, Sire, and note the lever by your knee."

The King obliged, spotting both His Home Secretary and Foreign Minister struggling to reach the top of a large staircase, reached down and tugged the lever causing the stairs to flatten forming a helter-skelter, sending both men sliding to the ground one atop the other.

"Dear Henry, you do me proud – I have not laughed this much for far too long."

"Then may I suggest you save some for my greatest trick?"

The King was now like a puppy, happy to be led wherever Henry deemed fit, "Does Chancellor Ernle please you, Your Majesty?"

"Sometimes, why?"

"Is he deserving of some mischief?"

"Most definitely."

"Then I hope this will please you, for it will most definitely *not* please him."

As the King looked on intrigued, Henry led Chancellor Ernle to a particularly comfortable-looking chair into which he sank with a suitably comfortable sigh,

"Sir, His Majesty informs me that you have served him well, and now the time is right for you to take a back seat for a while."

On a nod from Henry the King pulled the lever setting Henry's greatest trick into motion: metal bracelets revolved around His Chancellor's wrists pinning him to its frame.

Charles laughed aloud for all to hear, "You need to learn to relax, Chancellor".

"If you say so, Your Majesty."

A second pull on the lever sent Chancellor Ernle in the reverse direction along metal rails winding through the rear doors, crossing the lawn at high speed before ejecting the terrified man into a fast-flowing stream powered by the Windmill.

"Mind your head, Sire."

At Henry's alert The King looked up and gasped as another of his MPs burst out of an upstairs window vomiting as he was deposited in a tree.

Henry performed a deep bow, "I do hope my show is pleasing Your Majesty?"

The stream now had two men in its grasp, each one struggling to stand upright and, most importantly for the King, each looking ridiculous to the amused onlookers.

Samuel Pepys endeavoured to help the sodden men out of the water despite doubling up in agony himself, even dropping into the water up to his chest as a result.

"Observe, Henry – there is but one man amongst all the confusion willing to bear discomfort to aid his fellow man – despite most of his fellow men being unworthy and Mr Pepys' bladder being most insistent."

"So that is Samuel Pepys - is his confidence in the water from being an Admiral?"

"No."

"A superlative sailor?

"No."

"No?"

"No, but he is a great Administrator. I have employed him in that capacity for my Royal Navy. Moreover, Mr Pepys is a man of character – a quality I value most highly. He was the first to inform me of the Fire, and the first to advise the felling of houses to stem its progress. I am forever in his debt."

"I am told he admits to being human, encompassing all the frailties of man."

"Rightly so. I am aware of his infidelities, but I am in no position to judge him on that account, yet, as for judging his character, I can state with complete honesty my pride in employing his services, for there is no-one better suited to his position than Mr Samuel Pepys…as one day, I hope I shall also be able to say about you."

"I shall not disappoint you, Your Majesty."

The King paused for a moment to great effect, for it made Henry pay even more attention than he would have otherwise, to His following words,

"If truth be told, I was hoping your father was of a higher calibre."

Henry rushed to his own defence, "Please do not judge me by his standards, Sire."

"Have no fear on that account, I shall judge you purely on your own merits, as I do with all my subjects."

"Forgive my impertinence, Your Majesty, but did I please you this day?"

"You made me laugh, and for that I am grateful."

"Then would you regard me as highly as Mr Pepys?"

"Do not push too hard, Henry, or you may push me away. Mr Pepys is special indeed, because he is noble, a quality most people value. Even a murderer on the gallows would try to remember a noble achievement to barter with at Judgement Day. However, what you possess is a light touch, not to be underestimated."

As if by illustration, one of the Ministers was embracing a maid in a nearby arbour when it suddenly sprang shut, enclosing them like an exotic plant before tipping them into the water. The King laughed heartily, patting Henry on the back,

"You have achieved two great things this day, Henry - publicly humiliating those public figures, and making your King laugh. You have tickled my funny bone today, young Henry."

The King set off, beside himself with joy, leaving Henry feeling as hollow as his old family tree after the lightning strike, the mention of the ulnar again sinking him so...

As the sun began to set and the MPs climbed vomiting into their carts, both Samuel Pepys and Tony Tyler recorded it all with quivering quills, concocting their own versions of the day's events for the benefit of their readers.

And so, from that day forth Henry felt under pressure to ensure the King considered him as noble a man as Samuel Pepys, and in so doing, the King had proved his keen understanding of people, for the best way to retain a subject's loyalty is to keep him attempting to impress...

CHAPTER 31
ABSENCE

It is easy to become so overcome by huge favourable events that smaller ones that carry more meaning get overlooked.

So, as Henry readied himself for another exciting venture for The King of England, his Business Manager was managing business at home as usual - selling her husband's self-portraits, playing cards and engravings.

So, when the time came for Henry to board the King's Coach for London, he was faced with a matter of conscience, and also the matter of his wife...

"You have made a healthy profit and moreover The King doth seem greatly pleased."

"And so He should be – I have humiliated many of those He despises.,."

"...and I have made a healthy profit for us to boot. I have sold almost everything."

"Then I shall depart a richer man."

Henry was about to climb on board, when his wife used the only tool at her disposal to delay him,

"...Yet, you have failed to mention me once in all of this, my love."

It had the desired effect, causing her husband to pause and consider his position for a moment.

"Really? Then I apologise, in all the excitement I must have forgotten. Will you not come with me to London?"

"I have too many practical things to arrange here - but you must go, Henry – I would only slow you down. I need to manage the repairs."

"Well, only if you are sure."

The Royal Coach jolted as it turned to depart, followed by the carts leaving in the opposite direction transporting the vomiting Ministers.

Henry turned to face his wife, but in so doing proved he could not face her fully, "The King is insisting I join Him."

"Then so you must. Do not worry – everything will be sorted by the time you return."

"It would be miraculous if you can achieve all you say in just two days."

"Have no fear – I have made a promise, and I intend to keep it."

But when Henry climbed aboard the coach, all the exhaustion of the last few weeks got the better of Jane leaving her as vulnerable as a hatchling bird, and as you know where there are hatchling birds there are also the inevitable predators close by, waiting to take advantage of the situation - such that as Henry's coach departed, it was as if all of her spirit had departed too, and as Jane stumbled, an arm appeared from nowhere to catch her. At that very moment, for whatever reason, Henry was looking out of the coach and caught sight of his wife in the arms of another man - the delighted Jacob Wiley, eager to profit from the situation.

Henry felt the sad taste of inevitability, yet no bitterness towards Jane for he knew he had neglected her far too often of late.

However, his thoughts of Wiley were far less forgiving, and he resolved not to interfere, but bide his time and see how things played out...

He was right to be wary of Wiley, who was already making mischief seeking profit from being alone with his wife, "So, Henry has gone to party with the King. I am certain he will remain a gentleman in the clubs of Mayfair, amongst all those pretty temptations."

"We actually both agreed it would be best that I stay to manage the house -"

At that precise moment one of the servants called,

"Madam, I believe one of the dunking seats may be damaged."

As Jane was led away, Wiley pressed home his advantage,

"Henry has worked so tirelessly on your House, he deserves all the wine, food and parties the King can muster."

Jane watched Wiley departing with distaste, yet despite herself his words had taken effect, and she could not avoid a twinge of concern…

CHAPTER 32

LONDON

Having never visited the delights of Mayfair or indeed ventured into a gentleman's club before, Henry was lost at sea – an ocean of extravagantly attired guests flowing through every door – ladies with painted faces displaying whalebone-supported bosoms, amongst other members equally painted around the lips and eyes, but sprouting hair on their chins and attired even more extravagantly than the ladies.

To Henry's horror, the King beckoned from the other side of the room compelling him to wade through this variety of socialites, aware of their fingers poking and prodding him like diners testing a passing delight. The King came to his aid,

"The Molly Room is on your left - do not enter unless you know yourself well."

"What do you mean by Molly Room, Sire?"

"It is a safe place provided for those who prefer to be called Molly than Monty - if you get my drift."

The look back from Henry indicated he did not.

"They dock in a different port, so to speak."

But the look back from Henry indicated he still did not.

"Molly is a girl, Monty is not - there are those who live adrift between girl and 'not' and who need a safe harbour to dock, for there are many who would wish them harm. As far as I am concerned, I know who excites me and where my ship will always dock."

To Henry's surprise, the King shook an unusual bottle of wine until it ejaculated froth,

"It is time for you to taste the rewards of wealth, Henry. This new wine from Cousin Louis personifies a great collaboration between our two nations – French flat wine brought to life by English bubbles. Permit me to introduce 'Champagne', which I deem should be enjoyed at every social occasion."

And with that the King released the cork with a loud bang bringing gasps, shrieks and young ladies aplenty suddenly, appearing from every shadow proffering their glasses.

A harpsichord struck up a familiar tune of the day, prompting the King to lead two pretty ladies on a merry time of twirling and kissing moments, before Henry recognised a famous face entering the room – famous for being not only pretty, but the apple of His Majesty's eye – the face he had restored on the locket. As soon as the King noticed Nell Gwynn, he pulled away from the two ladies spouting dutiful excuses to join her,

"My darling Nell, how wonderful, I was just perfecting my kisses for you. I have sheets of silk, and pillows of swans' down awaiting our pleasure. Pray, come..."

Now suddenly alone, Henry scanned the room for a friendly face but, being in unfamiliar waters and under the increasing influence of alcohol, his gaze continually came to rest on the undulating bosoms...

As Champagne overwhelmed his reasoning, and his eyes kept wandering from bosom to buttock, Henry's lungs became acquainted with the floral swirl of Opium, having been exhaled by most of those present, flooding the room with its intoxicating aromas inevitably reaching Henry's nostrils, seeping into his lungs, and making their way to his brain where their influences went to work. Henry became detached from all that was real, fancying he was witnessing the strangest of things – as if the music was compelling dancers to lose their clothes before his eyes, releasing all restrictions from their swaying hips...

56 miles due North, Jane's hips were swaying too, carrying her frame unsteadily towards the living room, until they gave way causing her to fall into Jacob Wiley's arms once more, and very uncomfortable to be so.

"Why are you here? Unhand me."

Her anger unlocked Wiley's sarcasm for a rare instant, "Please do not attempt to thank me for saving you, I am certain your husband would have done exactly the same…if he were here, and not partying at the King's pleasure…"

With anger the only driving passion left in her, Jane used it to the full, pushing Wiley out of the door sending him crashing into his room.

Coincidentally, at exactly the same moment, 56 miles due South, Henry was navigating his way to the men's room…

…meanwhile 56 miles due North we find Jane arriving at her bed before sinking into its silken sheets, whilst a mere 20 feet away, 1 floor below Wiley, sensing his opportunity, snaked across the floor to her room, finally reaching the door…

At that precise moment, 56 miles South in London, Henry opened the toilet door and jumped at the sight of semi-naked dancers dabbing rice powder onto their white lead cheeks causing deadly small explosions of delight...

At that very same moment, Jacob Wiley came to a decision - be brave and nudge the door open, which he did revealing Jane in all her semi-naked glory, supine across the bed. In each location, both men remained staring for a few moments, each considering their options.

The dancers smiled at Henry's innocence as he sat stiffly, his eyes having never witnessed bare breasts in such numbers before. Eventually a dancer named Rose twirled over to him, leaving Henry's head to cope with how gravity could compel each bosom to sway independent of the other under the influence of her perfect hips. Not knowing what to do with his hands, Henry felt it best to do nothing, leaving them to comfort his own body – for after all, they had been the closest of companions these last 36 years. Everything came to an abrupt halt once Rose took hold of his hand, and placed a pipe in his palm, curling his fingers around its stem. Henry watched mesmerised as her full, red lips parted to accept the stem, and began to suck on it in a most seductive manner, attracting more girls to gather and watch. His nerves were such that when Rose withdrew the stem from her lips and pressed it against his, he became as helpless as a baby, sucking at its opioid vapours until he exploded into a fit of coughing causing his eyes to burst, much to the amusement of the ladies. He felt such a fool, but the ridicule he had expected never arrived – only whispers of encouragement flowed from those lips that set about kissing his cheeks, bringing Henry fond memories of being cradled in his mother's arms so many years ago.

Meanwhile Wiley's lips had arrived within inches of Jane's, but they held no interest for him save for her gold fillings tantalisingly only inches away - the temptation was too much -

He reached -

She awoke -

He gasped -

She bit -

He screeched -

Her narrowing pupils were still feeding his image to a mind not yet fully awake.

Both mouths gasped as one, before both sets of lips set about wildly wriggling, each mouth attempting to formulate a meaningful sentence,

"What are you doing, Mr Wiley?"

"…Why, protecting you of course."

"How so?"

"I heard a sharp noise from the drawing room moments ago, and upon realising you were alone, raced up the stairs to check if all was well with you, in the same manner that I always consider your well-being, Madam."

"Really? Is my house truly empty then?"

"Quite empty, Madam. My only concern was for your welfare, now that your husband is away enjoying London."

"My husband will be working hard to promote our business, nothing more."

At that moment both mouths paused considering their next move, especially as one was at the mercy of the other hovering so close above, she could taste his breath.

Within seconds their impasse was resolved, and quite innocently, by Jack,

"Shall I check for breakages, Jane, or can I return to my dear wife who will be wondering where I might be?"

Jane spotted Wiley's eyes shifting wildly above, as his mind tried to cope with the situation.

Thoughtfully, she came to his rescue, "You may go for the night now, Jack - and thank you."

Jacob Wiley was now left in an awkward position, straddling the semi-naked wife of his employer who was miles away, and unlikely to return any time soon. He heard Jack close the door, then faltered, considering his options, for there were only two people in the House now, each watching the other with a hint of expectation,

"So, Mr Wiley, what are we to do?"

Wiley faltered – Jack had just referred to Jane by name, and she had returned the compliment, but that courtesy was still being denied him...

The thought troubled Wiley to such an extent his eyes broke contact, and he slipped away heading for the sanctuary of his own room one floor below, closing the door with a sigh of relief, protecting his personal space whilst one floor above, the woman who held his future in her hands, remained silent.

Wiley began to panic – he could have so easily taken advantage of Henry's wife, yet something had stopped him. And then a thought occurred to him - why hadn't Jane called on Jack for help?

Wiley kept asking himself that same question as he paced his room just one floor below – and the more he circled his empty bed the more he imagined Jane lying upon it. Her face plagued his mind – the way she had stared at him with that hint of expectation, completely at his mercy...

Wiley spent the night in torment listening for Jane up above, and having heard nothing for hours, began to wonder if she was safe...

And that was how it happened - for the very first time in his life, Jacob Wiley had cared for another human being, and it irritated him. He was furious - his boss's wife was an easy target but sympathy had held him back. He could not afford such feelings, for the world of finance held no room for sentiment - 'fortunes are lost during the blinks of doubt' - as his father used to say.

Jacob Wiley had placed money in the highest regard for the whole of his life, far above family and friendship, yet this magical house was offering him the chance of something far greater – the opportunity to be part of a family business where each member supports the others – a wholesome partnership.

And it terrified him.

The terrors continued for hours under the light of the moon until, unable to face any more, Jacob Wiley slid under his covers to face a battle he had to fight alone – between the man he wanted to be, and the person he actually was...

CHAPTER 33
CHANGES

T he House of Wonders was proving most lucrative with a reputation that had spread like the dreaded Plague, crossing county lines and country borders, and for many months Jane had managed the House alone, but exhaustion eventually took its toll, allowing responsibilities to be passed on to those who did not deserve them, such as Jacob Wiley ingratiating himself as her Manager of Finance (the position for which he had always aimed) by making Jane increasingly reliant on his advice in all things, no matter how unqualified.

Thankfully, Jack still bore an honest soul, assuming the role of House Manager exceedingly well, being responsible for the hiring of staff, plus the upkeep and security of the buildings. And it was in that capacity, whilst checking all the rooms early one morning, that he stopped at Jane's upon hearing voices from within. Through the partly open door, he spotted Wiley hovering over Jane with the sunken eyes of a hawk, having spent all the previous night spying on her private finances...

Jack knocked on the door sharply, ending their discussion in an instant,

"Good morning, Jane..."

Both heads turned to face him – Jane's wearing a smile and Wiley's not.

"Is there anything I can get for you, Madam?"

"A cup of tea would be lovely thank you, Jack."

Jacob winced - once again they had referred to each other by name and ignored his.

Jack returned with the tea and, more importantly, a copy of 'The London Gazette', which Jacob pounced upon as soon as it landed, like a seagull swooping for titbits. Suspicious of Wiley, Jack hovered in the background ready to intervene, yet Jane appeared to be perfectly relaxed, excited even.

"I wonder how Mr Tyler has reported our Grand Opening."

Wiley could not help sprinkling a pinch of sour onto this sweet mix,

"It is a great shame Henry cannot be here to read it with you, Jane."

Despite remaining in the shadows, Jack was keenly aware of Jacob Wiley's charming intent, such that when Jane laughed, he was relieved to discover it was purely due to that article, and nothing more.

However, on the verge of carrying out his duties, Jack's blood ran cold - Wiley was now congratulating Jane with a touch to the shoulder, and she was responding with a smile,

"Thank you, Jacob, Henry shall be very pleased! Come, let us welcome this day's visitors!"

It had not gone unnoticed to the delighted Jacob that Jane had called him by name – and it had not gone unnoticed by Jack too, hovering in the background…

As weeks passed into months, whenever Jane felt uncomfortable, Jacob Wiley always made sure he was close at hand, ready to act as her confidante. Jane was so grateful for his service, she addressed him once more by name, 'Thank you kindly, Jacob,' she had uttered one evening with such warmth, he had trembled from the surprise of it, thereby shaking off all pernicious thoughts, rendering himself as helpless as a child before her, for Wiley believed he was transforming into a better person entirely due to Jane - improvements he now wanted to maintain at all cost, for if he wished to retain her respect as Finance Manager, then would have to manage all matters of finance in a respectable manner, as it were…

Yet, as mentioned earlier, a person's true nature remains with them, and Wiley soon resorted to his old ways using cunning to access Jane's personal accounts, which to his disappointment were revealed to be in a healthy state.

Wiley immediately set about weakening the results, for it is only when deeply worried that one accepts advice from any person put forward as an expert, whether qualified or not.

By means of blag and blarney, Jacob Wiley succeeded in persuading Jane to sign an insurance policy late one evening, when she was too tired to concentrate, adding her signature to the offer of his 'Superior Building and Contents Policy' covering the House against the threats of Fire and Theft.

Wiley knew it would be hard to convince the Law of a theft on the scale required to make the payout worthwhile…but Fire was a different proposition altogether, only requiring a strategically placed spark to make the acquaintance of a suitably combustible material such as gunpowder, *(which coincidentally, the House held in abundance stored for its daily firework displays)* resulting in a huge payout, for which Jacob Wiley would receive a healthy percentage for his recommendation.

Wiley's eyes narrowed as he set about planning his next moves, and not for the first time in his life, Jacob Wiley felt shame.

CHAPTER 34
AN UNREAL REALITY

As Henry's absence had grown from weeks to months, it should come as no surprise that changes had taken place in the business: his playing cards were still selling well, but it was his House of Wonders that was proving the most profitable, attracting visitors in increasing numbers as its reputation spread like tea on a towel across the English Channel, providing Northern Europe with a pleasant distraction from its Wars, Corruption, Disease and Famine.

Using a purse grown fat on those proceeds, Henry chose to invest in 'Respectability' with the proud purchase of two ships ('Constant' and 'Snowdrop') and frequenting a new type of establishment called a 'Coffee House' on London's Piccadilly - a place provided by Mr Edward Lloyd for underwriters to mingle, whilst spilling their secrets encouraged by the free offerings of coffee.

It was in this most civilised of places where Henry attempted to liberate men of money from some of their wealth, to invest in his most fantastical idea to date – a magical attraction to be built in the very heart of London, based upon an idea he had expressed a long time ago to the landlord of 'The Golden Goose Inne' - if only he could remember clearly what it was, for opium had immersed him in a dark world where he was losing his grip on reality.

Back in the Mayfair Club, the confident Henry of old was no more, as we find him shakily negotiating its deeply piled corridors, a broken man, unable to remember much of the last few months as if they had been dissolved in acid through the influences of alcohol and those opioid vapours...

As he wandered the rose-scented corridors, Henry's mind was facing troublesome questions – had he performed anything untoward these last few months? He could not remember and feared the worst – not helped by another male guest grinning conspiratorially and adding a wink as he passed. Then a sudden flash of images startled him: – Rose's red lips sucking smoke as her hips rose and fell, as if aboard a boat riding the waves...

Rounding the next corner Henry was taken aback, stumbling to his knees upon colliding into another bewigged gentleman who crashed to the floor just as he had, smashing into the carpet inches from his face.

Gazing up at the wretched fellow Henry was transfixed by the sorry sight of a man of similar age, drained of all enthusiasm with bloodshot eyes peering out of what appeared to be the pale face of a clown riddled with spots.

As Henry staggered to his feet, the wretch did likewise and as he reached for support the wretch did too, their hands meeting on a wall of glass framed in decorative gold.

Unable to stare himself straight in the eye, Henry moved off, heading for the stairs aiming to descend back into the real world and hopefully find himself again. In his desperation to escape the club's scented stench, Henry stumbled for the stairs, descending its spiral - his shoes slipping on the carpet's pile as he passed mirror after mirror, each confirming his descent of mind and body into the wreck who joined him on the ground floor and accompanied him outside.

As he staggered from the club into the hustle bustle of Piccadilly, Henry felt relief - he was entering the real world once more, immersing himself in the chilled air much like his refreshing swims in the River Cam as a child, but as he waded through waves of busy folk charging to left and right, Henry caught sight of his reflection distorting along each shop window, like a sceptre walking in a 'half-world' of its own, eventually catching sight of something tangible to aim for - in this case the safe haven of the Lloyd's Coffee House looming 50 yards ahead. Henry breathed a sigh of relief and headed for it, opening its doors into a calm sanctum of familiar sounds and smells - the aroma of coffee providing a welcome change from the fragrances of the club, and the calm drone of gentlemen's discourse replacing the indiscreet giggling of young ladies.

Sinking into a comfortable armchair Henry was immersed in tales of adventure all around him spun by captains having returned from exotic lands firing his imagination with tales of fiery dragons, snapping crocodiles and slithering serpents. And as he sipped the coffee, it was as if a fog was clearing from his mind, returning details the opium had filtered away including the mechanics of his next wonder which, to his great relief, gradually returned in their entirety.

Yet, practical considerations still remained: of where and how it should be staged so he waded into the ebb and flow of Piccadilly with its anonymous hustle and bustle, heading West over cobbles passing imposing buildings of stone towards the natural beauty of Hyde Park Corner and its trees.

As he got closer, dodging the coaches passing to and fro and horsemen competing in 'riding in the ring'*, Henry fancied he could hear the sound of a distant organ, and as he was drawn to the sound he noticed others had been drawn too, growing more numerous after each step, and of all ages from those bouncing a ball to those walking a hoop, either on foot or saddle or in carriage, converging on either side of him irresistibly drawn ever faster until even Henry began to jog, caught in the thrill of it - and suddenly there it was, magnificent amongst the trees like a massive flower whose skirts were being lifted by folk swarming inside like bees drawn to pollen. Henry ventured inside too, leaving the ordinary world behind and became enveloped in a most wonderful place of entrapment - of heat, sounds and smells only apparent once inside its cocoon - the rank odour of caged beasts, plus an enormous elephant towering over its tiny trainer attempting to lead it with a rope and stick, but most impressive of all to Henry's mind and too bizarre to comprehend - the blasts of fire from the mouths of clowns.

Henry took it all in with eyes ablaze, desperate to return to his studio to annotate everything whilst it remained freshly trapped in his mind.

Back in Littlebury, upon hearing a horse galloping toward the House, Wiley burst from his room intercepting the messenger at the door, swiftly relieving him of the London Gazette, sneaking it up to his room and rifling through the pages until finding the article he was seeking, with a headline that made him snort with glee:

'Mr. HENRY WINSTANLEY DISCOVERED RESIDING IN LONDON'S PICCADILLY AT THE KING'S PLEASURE.'

Wiley could not suppress his joy, reading the article aloud with much amusement,

"It has come to our attention that famed showman, Mr Henry Winstanley of Littlebury, has been residing at His Majesty's pleasure in London's Piccadilly, quite literally it would seem, for his lifestyle has emulated the King in respect of being entertained by the company of beautiful ladies…"

Beside himself with glee, Wiley propped the Gazette against Jane's door and retreated, unable to suppress the occasional snigger, as was his nature.

We find Jane also practising her true nature - managing the house, seated in her wheelchair by the turnstile collecting the first admissions of the day, when a pair of attractive young women eagerly approached, offering their entrance fees.

"Tell me – is Henry Winstanley in residence today?" - a frequent, yet not impertinent question to which Jane answered with professional detachment,

"I am afraid not – he is on business in London," compelling the woman to snatch her money back. The second young woman approached,

"Will Mr Henry be here tomorrow?"

Jane remained polite, "I am not certain of that, but I am certain that he shall be returning very soon."

Whereupon, the second woman also took her money back,

"Then *we* shall also be returning very soon."

Jane had no trouble hearing their banter as the young women departed chatting loud enough to be heard,

"I think that may be his wife."

"Poor thing – we have nothing to fear from her, then."

Despite her attempts to ignore them, that parting comment had touched a nerve, bringing a flush to Jane's cheek and a tear to her eye.

As was often the case of late, Wiley suddenly swooped close by as if waiting for such an opportunity, bringing Jane the comfort she needed. A splash and the accompanying shrieks of laughter from the gardens beyond, acknowledged the dunking seat in action.

"They are having such a merry time, Madam, and I hope Henry is too, whatever he is doing and whomever he is doing it with."

Jane smiled politely in acknowledgement, but Wiley sensed a sadness to her spirits which most certainly lifted his stating, "The weather is being kind enough for me to predict another profitable day. Are you seeing the same numbers we've enjoyed throughout the week?"

"Let me see," replied Jane with a playful tone that disappointed Wiley, for as he stood watching her sitting in Henry's chair, he knew that Jane's thoughts were already back with her husband as she toyed with her abacus, "Numbers are increasing slightly, thank you, Jacob. Visitors are coming from farther afield now – we even had a family from Ireland yesterday. Henry will be as pleased as you are, for lately, I have noticed increasing numbers of visitors carrying your insurance policies. Business must be good, but you still wear the same suit - my mother could help you in that respect, if you were to ask her…"

Jacob shook ever so slightly - so slightly in fact that Jane had no notion of it,

"My thanks, but I have ordered one from London, quite close to my office, as I was able to enjoy a small discount. And, on the subject of money, although your dexterity on the abacus is admirable, I could suggest ways for you to improve your own accounts…"

No sooner had the words come out than Wiley regretted them, for Jane swiftly adopted a quizzical expression and turned to face him with it…

"You have been looking at my personal accounts then…?"

Wrong footed by this woman again, Jacob stalled for a moment, searching for the words to get him off the hook, when another woman came to his defence in the shape of Jane's mother, stepping down from her carriage, with eyes busily scanning everything:

"My Goodness, Jane, I did not expect you would have so many visitors. Is it like this every day? If I were you, I would take advantage of the situation and place an advertisement for our shop right here - it wouldn't do for a mother to turn down an opportunity gifted by her own child, don't you agree, Mr Wiley!"

Wrong-footed by a woman again, Jacob could only shrug, which he performed to glum perfection.

Mrs Taylor was in no mood to delay, and swiftly pushed her daughter ahead,

"Come, Jane, I need to talk with you privately, we will have much to discuss these next few days, woman to woman - your father will welcome the break as much as I…"

And so, they departed, leaving Jack to manage the tickets, whilst Jane led her mother and all her bags up to the house, unashamedly pleased by her visit.

Curiosity getting the better of decorum, Jacob kept close behind, unnoticed by everyone except Jack managing the turnstile.

The Gazette was still lying untouched by Jane's door as Jacob scurried across the floor to catch their conversation. Jane and her mother were attempting to conduct a serious conversation, unaware that Jacob Wiley's earlobe was so close, pressed hard against the door,

"Do not ask me why, love, but our men suddenly have it in their minds to build. Before he went away, your husband suggested your father and I expand our business. He had some good ideas, but we do not possess a modicum of your wealth. Please help me in that regard, as I have always helped you."

"How much are you thinking of, mother?"

"Well, your father has no problem thinking big and 'big' means expensive. Will we have to beg at Mr Wiley's door?"

Upon hearing the mention of his name, Wiley moved even closer, despite there being scant room to do so, causing his ear lobe to throb – it being pressed so tightly against the door, causing pain for very little gain as what was being said became indecipherable. Even so, at least Wiley could tell the next voice belonged to Jane,

"Mr Wiley has not got that authority, mother."

"Then pray, why do you allow him a room here? I always trust my instincts, Jane, and that man is not all that he seems."

"I used to feel that way, mother, but now I'm not so sure. Henry likes him and…"

"Why do you let him reside so close? Business and Personal affairs should dwell under separate roofs."

Wiley's blood turned ice cold - after that condemnation there was only one fate allowed for Mrs Taylor of Taylor's The Tailor Shoppe - and it was not a favourable one.

Mrs Taylor continued, "I suddenly find myself needing the smallest room in the House – will I need to evade your husband's contraptions to find it…?"

Wiley slid away, as so often was the case, into the shadows, settling into a corner where he had a perfect vantage point of any comings and goings from Jane's room. Within seconds Wiley was rewarded with Mrs Taylor emerging to venture down the hall in search of the bathroom, wary of Henry's mechanical 'surprises' along the way. Jacob slid back to the part-open door and, spotting Jane turning away, seized his chance and propped the Gazette prominently against her seat, then beat a hasty retreat arriving at his door when a shriek froze him to the spot, coming from near the bathroom. In her journey along the landing Mrs Taylor had inadvertently stepped on a fake floorboard, triggering a bell that sent her running to the safety of the bathroom and locking herself in.

Jane let out a laugh, guessing what had occurred, and it was only then that she noticed The Gazette left propped against her chair by Jacob Wiley, who coincidentally had just arrived in his room one floor below scurrying across to the door, eager to hear anything from the floor above.

He heard nothing for a while, but that didn't mean nothing was happening, for Mrs Taylor had sunk to her knees in the lavatory, gasping by the sink, staring wide-eyed at a huge spider that had sprung over the basin beside her, now dangling over the edge. No sooner had Mrs Taylor realised its purpose, than the spider retreated to whence it came via a clockwork mechanism in readiness to replay its trick on another guest. Mrs Taylor cussed, then sighed with relief and waited patiently for nature to run its course.

One floor below, Jacob Wiley heard the sigh plus the tinkle of nature running its course, and moments later the pouring of water, followed by footsteps as she headed to the floor above his head and shuffled towards his room.

With beating heart, Jacob caught sight of her peering into his room and headed to intercept her.

Meanwhile, Jane had found the Gazette and began to turn the pages, unaware of her mother's predicament one floor below...

Mrs. Taylor was rummaging through Wiley's clothes, completely unaware of him approaching from behind when her eyes popped wide open upon finding Wiley's stolen jacket and the damage caused by the nail.

As she ran for the door Jacob leaped out from the shadows, grabbing at her necklace at the same time as one floor above Jane's eyes opened wide upon reading about Henry's alleged infidelity.

And so it was that on different floors mother and daughter gasped as one - Jane dropping the Gazette in disbelief as one floor below, her mother gasped for breath as her mouth was forced wide open to accept the necklace being pushed inside dislodging her fillings, until it jammed upon reaching her throat. Wiley struggled to keep her quiet as she snorted for air, scratching at anything she could reach. He managed to drag her out of his room onto the landing, horrified at the strength of this middle-aged woman flailing her arms about blindly, stabbing at his nostrils, ears, anything she could find - her eyes bulging fit to burst. He saw her spotting the rosary bead around his neck, and her eyes dilate as she remembered discovering it behind her shop. Jacob continued ramming the necklace down her throat struggling with the weight of her head whilst dodging her hands clawing at his face. Wiley caught himself in the landing's mirror and saw Satan staring straight back straddling the woman who was wriggling like an overturned beetle beneath him. Her eyes bore into his as her cheeks turned blue, the tip of her tongue emerging amongst the froth of saliva until she finally convulsed with a mighty groan and went limp - her eyes frozen as wide as her mouth which seemed to be locked in mid-scream. Wiley stared at the mouth in shock, expecting something to move, which eventually it did - a bubble of saliva sliding down into the abyss.

No cough – nothing at all.

It was Jacob who broke the silence, letting out a cry as he staggered back from shock and dived into the sanctity of his room, avoiding the windows, mirror, or any polished surface that could reflect his true nature.

One floor above Jane had come out of her room, attracted by the commotion, and looked down to see her mother looking back from below with the necklace stuck in her mouth. Jane dropped to the floor screaming, triggering a primitive response from her mother whose eyes suddenly banged open, and she grabbed at the necklace, trying to pull it from her throat. Wiley appeared at the door looking in disbelief at the woman he believed dead, now wriggling at his feet, gurgling as she tried pulling out the necklace.

Up above Jane jumped into action, struggling for the stairs all-the-while calling,

"Jacob will help you, mother!"

But Mrs Taylor was well aware of her situation, struggling to get away as her intended killer slid on top of her poking his fingers into her mouth, pretending to extract the necklace but forcing it further down. She bit down hard on his finger making him shriek, whilst all the time her daughter was sliding down closer, crying out,

"Jacob, please save her!" which attracted a reply as cold as a Billingsgate fish merchant selling fish,

"I am bloody trying to!"

Mrs Taylor's eyes were barely four inches from Wiley, and as he stared down willing the old woman to die her pupils widened, exposing a soul knowing exactly what he was trying to do.

"He's trying to help you, mother."

But Mrs Taylor knew different, her eyes resting accusingly on his as she felt her necklace chain edging down to the top of her lungs. Wiley continued forcing it down, willing her to die until her eyes stared into his and froze, chilling him to the core.

As Jane approached Wiley searched frantically for a hand mirror to push against her mother's mouth and confirm what he desperately hoped to be true, for Wiley was in no mood for surprises.

Jane collapsed at her mother's side utterly bereft as Wiley hovered alongside - fingers poised to stroke, poke or croak, depending on Mrs Taylor's situation.

Thud-thudthudthud - thud

The charge sounded like a legion of soldiers but was just Jack leading his team up the stairs to discover Jacob Wiley exactly where he wanted to be - cradling his Manageress alongside her deceased mother - one no longer posing a threat, and the other offering an opportunity of wealth, literally within his grasp.

Jack took control in the calm manner that had made him the ideal candidate for his position, instructing his team to remove the body with the due care its daughter deserved – privately vowing never to leave his Mistress alone with that man again, so suspicious was he of Jacob Wiley.

After a moment, Jane sat bolt upright, startling everyone present,

"Where is my husband?"

Jack provided the information as gently as he could,

"I am led to believe Henry is prone to staying at a gentleman's club in Piccadilly, Madam."

"Once again you seem to know my husband better than I. Take me to him and contact Mr Tyler - I wish an audience with both of them."

Then, as Wiley began to protest -

"Now."

Jack sprang into action at once, overtaking Wiley to collect the wheelchair, but Jane intervened so swiftly he released it,

"I shall manage without my husband's support thank you, Jack. Would you be kind enough to appoint your best assistant to manage the house whilst I am gone? - I trust your judgement above all on such matters. And kindly arrange a carriage."

Jack responded with his usual efficiency, escorting the distraught Jane to the gate as the carriage drew up, unaware Wiley was following a few steps behind - his mind reeling as to how best remedy this situation, for the idea of Jane rejoining Henry without him being present would be a disaster.

Spotting Wiley approaching, Jack swiftly helped Jane into the carriage before leaping in alongside believing he had therefore freed her from Wiley's clutches, but unbeknown to him Jacob had no intention of losing his prize and sprinted from the House catching up with the carriage, and managing to leap onto the back unseen by either occupant…

CHAPTER 35
A TIMELY MEETING

2 miles to the East of Hyde Park Corner lies a street that became the nascence of an area populated by - and therefore governed by - men of News, remaining so to this day – London's Fleet Street.

The character we shall now concentrate on is one we have met before in Saffron Walden - a shrew-like man with a voracious appetite for a newsworthy story, preferably one of sufficient truth to avoid being labelled untrustworthy whilst centred upon a character of sufficient intrigue to keep his readers hooked - (and if the story lacked intrigue, he would adapt the truth to ensure that it did.)

The man, you may remember, was Tony Tyler and as for the central character to his story - Henry Winstanley ticked all his boxes...

Jane entered Tony Tyler's office on the first floor a weakened woman, bearing not only a broken hip but the loss of her mother, and what felt like the loss of her husband too, so by the time she managed to sit down, Jane's strength had all but gone, making her doubly grateful for the support of trustworthy Jack.

Wiley arrived an hour later somewhat dishevelled and covered in dust - eliciting sympathy from Jane, but not from Jack who maintained a healthy suspicion throughout. A young runner arrived with glasses of gin for everyone present - Jacob Wiley seizing the opportunity to offer Jane his glass first which she accepted most gratefully.

Thus, the tables were turning in favour of Jacob Wiley, having won the support of the most influential, yet vulnerable person in the room who began the meeting,

"Please tell me, Mr Tyler, if what you have written about my husband is true?"

"I work for a newspaper, Madam - I am compelled to tell the truth."

Tyler's words hit Jane like a slap to the face, followed by a surprise…

"Your financial partner has kindly prepared the groundwork for this visit."

Jane hid her shock well and tried to remain composed using words aimed at Tyler but with eyes fixed upon Wiley, "Mr. Tyler I have stayed honourable to our verbal agreement, yet I read your article this morning regarding my husband's alleged extra-marital activities. Have you no loyalty to me?"

Tyler shifted in his seat, well aware of all the eyes fixed upon him,

"Madam, there is a price for fame in that one's activities become of great interest to the general public."

"Mr Tyler, please do not underestimate me, I fully understand the economics of your decision."

"Mrs. Winstanley, your husband's popularity with the public is to his great credit – a situation I was led to believe you were both striving to achieve. He has proven that huge profits can be made from the world of entertainment, so it is only natural that he would wish to create something bigger and better than before. In that respect, I believe his latest endeavour shall prove the most profitable of all."

At that point not only did all the eyes remain fixed upon Mr Tyler, but every eyebrow raised too. Most possibly, due to his insecurity, Jacob Wiley was the first to speak,

"Latest endeavour? What endeavour?"

Tyler replied as if everyone knew, for why would they not, "His 'Waterworks' of course – it sounds fantastical."

At that point, even Jack was united with Wiley and Jane - they had no idea at all what Henry was planning, prompting Jane to ask, "Mr Tyler, once again you seem to know more about my husband than his wife."

Feeling completely left out, Wiley interrupted quite clumsily,

"When does he plan to start building this Waterworks?"

"He already has. I have seen his sketches, and although I have no knowledge of engineering, the word that describes them best is 'fantastical'."

"That word sums up Henry very well," added Jack with a grin, making Jane feel more isolated than ever,

"I need privacy for a moment. Jacob, please take me to your office."

Jacob nearly fell off his seat with shock, "As I've already said - my office is being decorated, Madam."

Jane was perplexed, (as was happening quite often of late where Henry was concerned) so attempted to regain control by addressing the most empathetic, and therefore safest, member of the group,

"Jack, would you kindly find my husband and bring him to St James' Park. It is close by, and it is a fine day – the sun will brighten my mood."

Jack replied, as clumsily yet well-meaning as ever, "If I know Henry, he will be in the Lloyd's Coffee House. I hear it is quite the fashion these days."

Tyler was swift to deal the next blow, showing no mercy, "I disagree - unless I am mistaken you will actually find him seeking comfort in 'The King's Arms' on Drury Lane."

Jane threw her hands up in despair, "Seeking comfort in The King's Arms! That sounds more like him. Jack, please inform Henry his wife needs to meet, if indeed he remembers who his wife actually is…"

CHAPTER 36
AT ODDS WITH HIS REPUTATION

*A*s Jack set off on an errand to find one man of invention, less than two miles away a Messenger was setting out to find another – a journey that would prove of vital importance, not only to our characters, but to every sea-faring nation in the World...

There is a well-worn fallacy that London is a dull damp place suffering from constant fog and rain, which tourists believed to be true and only Londoners knew better, for when spending a lifetime there, they had been blessed with morning dews sparkling like jewels carpeting Green Park, and the majestic sweep of its industrious River Thames.

However, the day the Messenger of our story alighted from the river at the appropriately named Water Lane, the sceptics were proved right - the sky was dark with clouds emptying their load onto folk below, speeding this way and that, dressed in heavy coats with shoulders hunched, to their intended destinations.

Picking up his pace, our Messenger raced around the corner and leapt across the various puddles to reach the sanctuary of Trinity House - a noble institution dedicated to the safety of shipping by means of its buoys and lighthouses.

He opened the door into a world awash with Maritime influence from the brass galleon clock on the wall to the desk of finest Jamaican mahogany shipped over by The East India Company to London Bridge Dock only a stone's throw away. An insistent 'dinging' from an upper floor accompanied the Messenger as he passed the desk manned by an ancient clerk whose only associations with the sea seemed to be a beard, a pipe, a brass bell and the bottle of rum hidden between his knees. The beard came into action, rising and falling as he spoke, "You are late, sir."

"It is raining cats and dogs out there."

"I understand. Be careful."

The 'dings' grew louder as the Messenger sprang up the staircase between sea creatures sculpted into its balustrades, until reaching the door at the top with its brass plaque displaying the name of the man, he was due to have met 6 minutes ago - 'Mr. Samuel Pepys - Master of Trinity House'.

He knocked quietly, (which considering the insistent alarm made no sense at all) and set foot into a room of Maritime Pride endorsed by paintings of historic Naval importance placed with pride across its walls. The alarm was at its peak when he approached an important looking desk, where an important looking gentleman was embroiled in a mountain of important looking paperwork.

Upon handing over the top page, duly signed with a small envelope attached, Mr Pepys sighed with relief at the weight lifted from his shoulders, and adjusted his pocket watch to silence the deafening alarm, thereby symbolising the control Man holds over Mechanics.

"Have you been told who needs to receive this?"

"I have, sir."

"And do you know what to do if he shows reluctance?"

"Yes, sir - in that case I should draw his attention to the small envelope attached."

"Indeed, and I expect you know where to find him?"

"Yes, sir, I do."

"Excellent, then our business here is done."

And with that the Messenger span on his heels, and raced back down the stairs, crossing the reception where the ancient clerk's beard rose and fell again as he spoke,

"Please be careful, young sir."

The Messenger had reached the door and was in no mood to stop, but as the comment had been delivered with such warmth, he reluctantly paused,

"I beg your pardon?"

"Please be careful, sir. The River is dangerous when she is full."

But Timothy Stamp was a Messenger in a hurry, preferring to deliver letters than waste time on idle chit chat. He was about to speed away as quickly as he had run into the Courier's Office a year and a half ago, spent six and a half minutes on the interview then raced away as the newest employee of 'SWIFT COURIERS LTD'. Yet Timothy Stamp was also a decent soul and decided that spending a few seconds with this old man would do no harm - after all, he could always sprint a few yards later to make up the time so added, 'I thank you, good sir, and the same to you,' before speeding off again heading West from that place of distinction to another - St Paul's Cathedral, born from the mind of the genius he had been sent to meet, now surrounded by a group of Architects, the boldest of which had little to say except offer praise, as so many of his kind had done before,

"You are to be congratulated again, Mr Wren, on your finest achievement to date – I wager this Cathedral shall serve as an iconic symbol of our City for decades to come."

The ice having been broken another Architect felt brave enough to add his praise to the mix, "From this day forth I can tell my students I have shaken the hand of Europe's greatest living Architect."

As one would expect, Wren accepted the flattery with good grace, having experienced so much of it lately, and so it was in the same vein that he accepted the letter passed by Timothy Stamp, the Messenger,

"Who has written this?"

"Another eulogy no doubt," chirped one of the Architects good-humouredly.

Timothy Stamp ended all speculation, "Mr Samuel Pepys, sir".

At which point Christopher Wren took the matter more seriously, "I see."

Upon opening the letter Wren's expression changed as the weight of its contents drew him to conclude the meeting, "Gentlemen, I have been requested to return to the Drawing Board," which drew the inevitable comments:

"Surely not another church, sir," said one.

"Aye, are 52 not enough?" added another.

Wren accepted the remarks with good grace, and delivered his reply in the same manner, "I am requested to design a folly, no less!"

"Sir, the folly is in expecting you to waste your time on it."

The group's laughter threatened to obliterate the Messenger's next remark,

"Sir, Mr Pepys has asked that you open the small envelope before deciding on a reply."

Intrigued, Wren opened the small envelope, his expression changing the moment he finished reading the note, emitting a grunt at odds with his bearing,

"Inform Mr Pepys that I shall give the matter the attention it deserves - and that is all I shall say on the matter."

His task complete, Timothy Stamp happily sped away, unaware of Wren behind him discarding the note into his waste bin.

CHAPTER 37

OF PELICANS AND RECONCILIATION

*O*ur chapter begins in the Royal setting of St James' Park, as instructed by Jane to be the most suitable place for holding a reunion with her husband in the hope of reconciliation,

As Jane somewhat nervously, Jacob Wiley ensured he was alongside to supply advice,

"Are you sure this meeting is worth the wait, Madam? Those approaching clouds are plump with rain, making me concerned for your health."

"Rest assured; whatever I may suffer I shall make sure my husband suffers too."

"Yes, but you have suffered too much already."

"Let me be the judge of that, Mr Wiley."

Wiley backed away, in recognition of his position, but could not help adding a note of doubt for her to mull over, "I only hope he acknowledges your worth as much as I do..."

"Mr Wiley, I believe I know what you are trying to do and let me add this note for you to think upon - I bestowed upon you the title 'Manager of Financial Affairs' - and that is as far as I am willing to go with you on the subject of 'Affairs'."

Wiley looked as shocked as if he had been slapped in the face, which in some degree he had. So, having been rendered speechless that was how he chose to depart, sulking silently into the distance.

Thus, as the rain clouds loomed on the horizon, Jane waited for a battle to commence, determined not to show her husband any sign of weakness.

The time passed slowly, and despite her wealth, Jane chose to remain without creature comforts, so determined was she to face her husband in the rawest of moods.

London was unfamiliar to her, and she had to concentrate hard not to be distracted by the clattering of coaches passing along nearby Piccadilly, or the chatter of well-dressed folk going about their business paying absolutely no attention to a frail-looking lady with a stern countenance wrapped in a shawl.

Jane's anger remained steady and firm, yet as the minutes passed as surely as those well-dressed folk going about their daily business, she was grateful for the anonymity the city provided - a world away from the small-town nosiness of Saffron Walden. And thoughts of Saffron Walden inevitably led to thoughts of her mother, and how much she would have valued her maternal advice at that moment, for it had been given many times before and on many subjects, whether she had asked for it or not - on clothes, or business or the young man of enthusiasm who had declared, 'Madam, this delightful young lady has guided me to my destination when my intention was to guide the young lady to hers.'

The memory brought a smile, and the smile brought a change of heart and a hope that Henry would still show some interest in her.

Meanwhile, Jack was following her orders heading into 'The King's Arms' on Drury Lane to discover a somewhat tipsy Henry surrounded by a group of admirers, in a similar fashion to Wren, apart from the fact these were not Architects but young women all benefitting from his generosity at the bar.

Henry was surprised to see his old friend, and very happy to be so, "Jack! Join us for a drink - Ladies, I am indebted to this young man for introducing me to the joys of alcohol many moons ago, of which I have now become somewhat of an expert!"

The girls giggled spontaneously as giggling girls do, much to Jack's annoyance,

"Stop behaving like a fool, Henry, and come with me at once."

Being caught off guard in front of the girls rendered Henry furious, "I beg your pardon!"

The landlord was swift to interject, "Mr. Winstanley sir, do you require any help?"

Henry span around to face Jack, "Well, Jack - what do you think - does Mr Winstanley require any help?"

"Yes, I believe you need all the help you can get, Henry."

Henry took Jack to one side more swiftly than one would have thought, given the amount of alcohol he'd consumed, "Did you not hear the Landlord, Jack – I am not just Henry but *Mr* Winstanley. You must treat me with the respect I deserve, or I shall have you accompanied from this establishment."

"Have no fear, I shall leave behind both this establishment, and the man I used to call a friend. I shall return to the person who sent me here. You are hurting her, Henry!"

Henry's face sank at the thought, "She is in London? Why?"

"I heard her say 'His Majesty is losing a loyal subject to the allure of money'. The loss does not only lie with the King."

The clouds that had once promised rain were now delivering on that threat. Ever the pragmatist, Jane put all hope aside, stabbed her stick to the ground, and hauled herself up to her feet only to stumble with a cry knowing the damage it would bring, yet as she tumbled, a hand gripped onto hers, providing care then an arm supported hers and when she looked up to see who was providing such help, the face looking down was the one she had fallen in love with, a lifetime ago in Saffron Walden now showing the effects of the time in between.

As the raindrops wet her forehead, his kiss dried them bringing a gasp, his second kiss dried her cheek bringing a sigh and the many that followed brought laughter until all doubt had washed away, and she faced her husband as any wife would: she criticised his pallor, he criticised her frown, she poked his ribs, so he took hold and pulled her back to him.

From 200 yards away it appeared to Wiley as if they were fighting, so with a lightness of step he headed for the nearest bar and ordered a bottle of their finest brandy.

Yet, although they had been locked together, Wiley had misjudged their intent.

"Hello again, wife!"

"Have you really come back to me, Henry? As close as you may appear to me now, I know the second you begin planning your next wonder I will have lost you again. I must confess it would be easier to cope with losing you to a death you had not planned to some other thing that you had."

"There is no need for either - I have returned. Do you remember me saying 'I have in my head such wonders the world will marvel at how they could all come from just one head?"

"I do remember, Henry, but I also remember much more and lately I increasingly see a shadow where I once saw a spark."

"I am the same man standing beside you now as then. I shall soon reveal my new wonder, surpassing all others, created from all the delights that have inspired me over the years - from the rotating planets of my youth to the automatons of France, the fountains of Versailles and the circus over there!"

Jane followed his finger finding the huge tent standing proud behind the curtain of rain.

"That tent contains the elements of my new and greatest Wonder which I shall build over there!"

Jane's eyes followed his finger but could only see a line of stone buildings, "I cannot see a circus."

"You soon will, underneath my windmill."

"I cannot see a windmill either."

"I see it here," Henry said pointing at his head excitedly, "I still believe I am here to create something of consequence - this must be it."

"You have created enough already - now settle down."

"I cannot settle just yet. Once I find my true purpose on earth, I promise I will do as you wish."

Jane studied her husband's face so on fire and dearly hoped it could come true for him.

Henry took her by the hand as he used to when they were courting,

"Let me show you something wonderful."

Jane walked with him delighted at their reunion, falling in love all over again and willing to follow wherever he led.

"We are fortunate to be blessed with the most superlative of Kings - He has made an impression upon me as good as any father could on a son. Over yonder is proof He has left an impression on others too, those much more important than I -"

Jane followed his finger to a group of huge misshapen birds with wings larger than she thought possible, landing with surprising grace upon a small island giving reason to their shape - their huge beaks as baskets heavy with fish bending their necks double from the weight.

Jane's face curled into a copy of the bird, "They are distasteful."

"They are magnificent - Pelicans - a gift from the Ambassador of Russia to our King."

"Are we at war with Russia then, for them to hate us so?"

Henry laughed so much Jane followed suit and together they headed for the gate, walking under its arch of stone into a far greater one suspended by many colours of refracting light, as discovered by Isaac Newton nearly 20 years ago and now being explained by Henry to his wife…

As the brandy drinkers amongst you will know - the shape of the glass is intended to trap the aroma…but it also distorts any image passing through to the extent that when Jacob Wiley toasted himself inside 'The King's Arms' shortly afterwards, he failed to recognise the shapes passing behind his brandy glass and the bubble glass windows beyond, and it was only when he stepped outside that Wiley got a clear view of Henry and Jane merrily boarding a distant coach that he realised how united they now were, and the extent of the task ahead of him to disunite them.

Dear Reader - as mentioned before, it is difficult to continually act against one's Nature, for try as hard as you may, the likelihood is you will remain just as before - proving the reason why it is called 'Nature'.

So, no matter how hard he tried to act otherwise, Jacob Wiley would always revert to his ways of old...

CHAPTER 38

A FOOLISH FOLLY

*S*amuel Pepys's letter had remained discarded in Mr Wren's waste bin for three whole days during which time others of less note had dropped on top hiding it from view - therefore it was a surprise, when on the fourth day the hand that had originally discarded it returned to retrieve it...

The letter journeyed beyond the decorative iron gates of St Paul's Cathedral all the way to Windsor where it was laid flat upon a drawing board alongside various quills, an inkwell and a pair of spectacles.

The Great Man read the note again, affording it the attention it deserved: and after a few minutes thought, opened the doors to his vast collection of books, consulted their spines and retrieved various examples pertaining to matters of Maritime, Meteorology and Geology.

The note remained open throughout, acting both as a reminder and a persuader as his mind continually tried drifting back to his Cathedral, and this important request which remarkably had remained legible despite its rag paper having absorbed the ink and all the intervening drips of tea and other spillages in between times:

'**My most honourable Sir Christopher Wren – we the members of Trinity House ask you to consider designing a sea-born structure nine miles South South-West of Plymouth, whose complex nature has not been attempted since Biblical times. Its success may well come to be judged as the Eighth Wonder of the World – a modern Pharos, if you will...**'

Wren paced back and forth for days at his Thames-side home contemplating the enormous challenge placed at his door - to supersede what Sostratus had achieved for the 20-year-old Alexander the Great 2000 years before, creating a guiding light spreading 40 miles in all directions and receiving a rightful place in the annals of History as a result.

As he walked alongside England's greatest river with its passing boats both large and small, Wren contemplated the weight of the challenge laid at his feet - to create the first lighthouse built entirely at sea for the world's foremost seafaring nation, was an irresistible challenge for any ego - but he already had fame, he already had a knighthood and when the practical challenges of building a structure miles out at sea came into consideration, Wren began to lose enthusiasm…plus when the high risk of failure entered the equation, Wren calculated it as one risk too far…

Thus, the note was returned to the bin for some other poor soul's consideration - if any poor soul could be found foolish enough to consider such a folly...

CHAPTER 39
A TRAGIC LOSS

Acting upon a rekindled spirit of unity, Henry and Jane returned to the 'House of Wonders' and immediately adopted their former positions - Jane at the front of house managing admissions, and Henry at the back in his studio finalising designs for the Waterworks, and all the while Jacob Wiley kept a low-profile in between like a spider poised at the centre of its web...

And for a while all was well - until the fateful day Henry was walking through St James' Park en-route to Hyde Park Corner when he chanced upon news so momentous, he would have expected to have heard it first from Tony Tyler's Gazette, not on the tongue of two passing scullery-maids...

"I 'ain't seen 'im walking them Spaniards for months…"

"Spaniels…"

"Yeah, I ain't seen 'im walking 'em around here for months so I knew it must have been serious - an' now I know why."

"What's 'appened then?"

"Well, I was up at the Palace when there was such a commotion - physicians runnin' this way an' that with pots of foul-smelling liquids donated by the King's own body as He tossed an' turned being bled."

"Sounds 'orrible."

"It was more than that - but you know what…despite it all, He turned to his brother and asked him to look after His poor Nell."

"Bless His heart."

"Not only was His brother James there but two priests, would you believe it, coming to drain what they could out of Him while 'e still had breath."

"What was it then?"

"I'm told it was an ulcer an' gout."

"I wish I could afford gout..."

"Aye, it's a rich man's displeasure, sure enough."

"Couldn't they have just let him pass in peace - that's what Priests are for, ain't they?"

"They hovered like crows till He declared himself a Catholic, at which point they fled like a gaggle o' geese, leaving the poor dear sucking at the air – an' all the while apologisin' for takin' so long to die! I felt so sorry for Him."

"Did you see Him die then?"

"No, but I heard Him as true as I'm standing 'ere. They are sayin' He died in peace but I'm tellin' you there weren't any peace to that dreadful shriek. I dared not look but couldn't help meself an' saw He'd left this world with mouth open as wide as Thor's Cave, God bless 'im."

"Was 'e still alive then?"

"Dead as a doorknob, just before all His bleedin' clocks struck noon like a load o' banshees, doin' me 'ead in."

"'Ow many of 'em?"

"Seven would you believe. I won't be sad to see the back o' that lot."

"D'you reckon they'll bury 'em along with 'im?"

"I'ope not, poor bugger - 'e deserves a good rest now God bless 'im."

"Aye, God bless 'im alright, we shall never see the likes of a better one…"

"I don't resent for one minute the sheets he soiled."

"Better boil 'em clean for the next poor sod then."

The laughs that followed were more like cackles, adding a commoner's touch - yet Henry didn't mind in the slightest because one of His Majesty's greatest strengths had been touching the common folk with His love of pretty ladies and fast horses.

The maids soon changed tack, eager to continue their gossip,

"Who's to be next then?"

"It's bound to be that dull brother of 'is - James."

"Oh, I do 'ope not - they say he's as grim as a sober judge…"

"Then God 'elp us all."

"…an' a devout Catholic."

"Ha! Then God 'elp Him!"

The maids departed chatting and cackling as if words were in limitless supply leaving Henry to endure the walk to Piccadilly alone and with heavy heart - abandoned by the man who had treated him more like a father than his own.

Wandering through the late King's Park that crisp February morning the barren trees stood perfectly still like a line of soldiers in mourning. Henry felt more alone than ever before - a murder of crows squawking the only requiem. And so it was that, while feeling adrift without a rock to grip, Henry came to a realisation - fellowship was the most important rock of all. Thoughts of Jane brought Love, thoughts of Jack brought Friendship and thoughts of Tyler brought Resolve, such that by the time he reached Piccadilly, Henry was refreshed and ready to pursue his goal once more - the new Wonder to surpass all others, with or without the dull new King's blessing.

51 miles North and equally alone, Jane remained loyal to her vow – supporting her husband's business interests - both the House and the continuing sale of his merchandise.

Words concerning fame travel far, even reaching King Louis' ears across the English Channel whose curiosity was sufficiently aroused to despatch a familiar figure to Littlebury - a dazzling Moor asking for Henry by name. He displayed African features wrapped in a rich dark skin clothed in exotic attire - exclusively black with silver trinkets. At first, Jane had felt insecure, for this Frenchman seemed extremely knowledgeable about her husband, yet over time he managed to win her over with generous helpings of Gallic charm.

Having no knowledge of French, Jane's response was initially modest yet friendly, commencing with an offer of tea, but she felt humbled by the sweetness of his response, decorated as it was with elaborate politeness in a voice of African descent flavoured with the arrogant seasoning of a Parisian, "Thank you, Madam, but my palette prefers coffee - it has so little experience of tea - a deficit which I am sure will be rectified soon."

Jane stumbled for a reply, "Bonjour, Sir, can I aid you?"

Feeling her response inadequate, Jane had to admit a great relief upon Jacob Wiley's timely appearance once again, greeting their French guest with a degree of French charm of which she thought him incapable,

"Comment est la sante de vos Rois monsieur?" (How is your King's health, sir?) asked in what Jane assumed to be a very good French accent, to which he replied in a perfect rendition of English, thus making a mockery of the whole thing.

"He is well, of course - my King always enjoys the rudest of health, especially with a close ally now sat upon your throne."

Omar Ricard turned to face Jane performing a discreet bow which, when added to his Gallic charm, pleased her greatly, "Permit me to introduce myself, Madam, – my name is Omar Ricard."

Jane replied smiling, "With a silent 'd'?" and adding upon his returning smile, "I remember Henry speaking of you with great affection, and, yes, I do admire your attire - my mother would have approved."

"My thanks, Madam. I do hope our friendship can also be enjoyed by our two nations, for France is your neighbour and friend. I am told King James is enlightened on the matter calling for an end to your English bravado. 'Do not aggravate, just capitulate and appease France', He says. Yet, what of Henri - how do you think he would react?"

Not expecting such directness, Jane was taken off guard and began with a laugh - not intending as an insult, more of creating a pause in order to think,

"Oh dear, you have me at a disadvantage, sir. Henry says your countrymen like to talk as incessantly of politics as we do about the weather. I confess to agreeing with my own countrymen on that score."

"But how do you think Henri would react? Would he comply with our laws do you think? We are most interested, now that he is your national hero."

"You flatter him - a hero of London maybe."

"A hero of England then."

"The rest of England may disagree, but I still know my husband well enough to say that he would definitely be in favour of a closer friendship with France, if it meant an increase in trade. He is a keen engraver which are selling well in England, and he told me of his eagerness to replicate that success in France with your grand houses... and his playing cards would prove popular at parties, and his painti-...."

M Ricard interrupted, but still in a gentlemanly manner,

"Henri is fortunate to have wed such a loyal wife. My King has heard much to admire concerning his achievements. I have travelled a long way to meet your husband again, could you say where I might find him?"

Jane laughed, "Ah, that is a complicated question, sir: Henry may well be a million miles away in some fantastical land of his own creation. I am afraid to admit I really do not know."

"But he enjoys good health?"

"I presume so."

"I see."

"I am glad you see, because I rarely do these days. His mind departs so often."

"Henri is a 'Grand Createur,' Madam."

"Ha, is that what you call it? I say Henry is a 'Grand Mystery' - even to me who should know him best of all."

A scream rang out followed by a great splash, signifying another victim of the dunking seat surprising only Ricard. Very soon afterwards, a group of visitors passed them laughing on their way out - some dripping wet, uniting both Jane and Ricard in a smile, which Omar elaborated upon,

"Henri has a way of bringing the child out of people - 'all the better to win them over', as he often said."

"I am glad you understand Henry, because I never really have - he keeps telling me his mind is like a young albatross constantly in flight. In fact, I am sure his mind is soaring in some distant sky as we speak - let me ask Jack to catch him for you -"

"There is no need, Madam, I will go and find Henri in my own way."

And so it was that with the passing of a coin, M. Ricard ventured towards the house to face all its trickery in search of Henry's genius for the curiosity of his King.

Jane turned the coin over in her hand, still warm from his grasp, until it came to rest revealing the head of the late King - a reminder of the loss of King Charles 14 months ago, made all the sadder by the coins still bearing his head in her hands.

"Henri is a very fortunate man."

Jane turned to face the owner of this new voice - a Frenchman of similar age to herself with a moustache that rose at each end of his lips, lifting them into a smile, adding a touch of youth to a face that was already bearing its age well.

"Madam, your gates have great possibilities."

"What a strange thing to say."

"I apologise; my English is not very good. What I meant to say is I could make your gates very beautiful."

"I am sorry, for my French is also not very good."

"Well, that suits Omar Ricard well, because he is not a very good Frenchman."

"Really? You know him then?"

"I know his sort."

"He seems quite forceful."

"That is the first reason I do not like him…"

"And he seems very loyal to your King."

"That is the second."

"You are most curious, sir."

"I find it helps to be curious - asking questions leads me to the truth - my work is governed by truth. I can make your gates the most beautiful in England, if you should allow me the privilege."

"Henry has never mentioned his dislike of them so I assume he must be happy."

"But more importantly, do *you* like them. Madam?"

"Fences are fences, and gates are gates. They control the bull from mixing with the cows. If mine regulate visitors during the day, and stop intruders at night, then they have served their purpose."

"And that is all?"

"That is all I need of them."

"Then I am disappointed. You are merely being practical."

"Of course, I am - what else is there to admire about gates?"

"Beauty, Madam! My gates transcend mere practicality."

"You sound like my husband."

"Then your husband and I are united in two things, at least."

"I am sorry?"

"An admiration for Practicality *and* of Beauty, both of which you own in equal measure, Madam."

What happened next took Jane by surprise by both its speed and compulsion, for she had no control over the blush that suddenly flooded her cheeks, displaying her feelings so readily before this stranger.

"You are most charming, sir, as is M. Ricard. It must be a characteristic of your race."

Jean Fitou's smile disappeared in a flash, shocking Jane both with its speed and the change of tone suddenly released - "Do not compare me to him, Madam. He and his King are at war with my faith."

"But He is at war with England."

"Then you and I are united, for He is at war on two fronts - with your country and His own countrymen who dare to follow a different faith. We Huguenots are being marginalised out of existence at the behest of the Catholics."

"I am sorry, sir, please forgive me. I had no idea."

"Ignorance is a consequence of living on an island. Yet, our countries are suffering from the same malaise - our faiths that preach forgiveness refuse to forgive. My Catholic King is determined to rid France of the Protestant scourge, of which I am proud to call myself a member..."

"Then please stay, sir, for I am sure I can persuade my husband of the need for new gates to adorn this house."

"My thanks, but there is no need for charity - my work in London is still months from completion. I am working for a hard taskmaster who demands all of my time. He is a genius though, so I am compelled to forgive him."

"Who is this man?"

"His name is Monsieur Wren."

"No! Christopher Wren?"

"Mais oui - you know him then?"

"I feel I know him quite well, through my husband. Henry has strong feelings about the man, and not exclusively of a positive nature."

"Good - then I have an excellent excuse to see you again."

Jane surprised herself with the boldness of what she suddenly proposed:

"Henry is planning a 'Wonder to surpass all others' he tells me - in London's Piccadilly. It may be something you would wish to collaborate on? I will ask Jack to introduce you, for your passions seem well suited to my husband's."

She beckoned Jack, warning him to avoid bumping into Omar Ricard for she had no wish to start another French war, especially on her own turf. As Jack led Jean Fitou away to find her husband, Jane was left feeling a surprising tinge of excitement at the possibility of meeting this Frenchman again on a more regular basis.

A group of visitors arrived, forcing her reluctant return to reality, and to her formal role as Manager of Affairs, yet within minutes the sound of squelching feet alerted her to M. Ricard approaching sopping wet from having made the acquaintance of the Ducking Chair, yet was in good humour nonetheless,

"I stand before you a happy victim of Henry's mischief."

"So, you found him then?"

"I found his wit and I found his mischief but alas, I did not find him."

Jane laughed again, "Monsieur Ricard, Henry is as elusive as a rainbow in the night sky."

"That may be, yet Henry's popularity is proving a popular topic of conversation within Versailles, and I can assure you nothing reverberates within those walls that does not eventually vibrate the eardrums of the King. This is the reason why I'm standing before you now, to relay your husband's popularity to my King, for every leader of men ultimately craves their support. I would remind you that although your countrymen may seem to some as being complacent about all but the weather, it is they who gathered to witness the axe severing your King's head, and my King does not wish the same fate to fall upon His. Popularity is the best deterrent to a Revolt, so in this context I shall report on your husband's popularity with the people of England, so my King can learn from it and ensure the people of France never have the hunger for a Revolution and the subsequent chopping of necks - He is very fond of his locks."

CHAPTER 40
OUT WITH THE OLD AND IN WITH THE NEW

T ime is a fickle fellow delivering constant change - the last three years had passed as if in an instant: the Merry Monarch had died passing the Crown to his dull brother James who, being a Catholic, was swiftly forced out tossing it back onto a Protestant's head - King William 111 of Orange, whose Princely wink at Audley End had made the young Henry blush all those years ago.

However, unlike Kings and their Crowns, the world of Finance is nervous of change, so today we find various Financial heads, including Insurance Underwriters, Merchants and Investors floating once again in their usual watering hole – Lloyds first Coffee House in London's Piccadilly, all familiar to one other, and all sat upon the same cushions they had squashed for so long, their posteriors had left impressions as vivid as the death masks gracing the nearby National Gallery.

Now however, they couldn't sit still, being so excited by the presence of a wealthy new acquaintance in their midst – Henry Winstanley Gent., recent owner of five new ships - aiming to secure trade routes across the Atlantic. And as usual when matters of business were being discussed over a coffee table, Jacob Wiley always made sure he was lurking in the shadows beneath, feeding off the crumbs they supplied...

On this occasion Henry was loving all the attention proclaiming: "My Waterworks shall require a huge thirst for tea and coffee, which my ships will quench with regular visits to the Americas. Indeed, as I speak 'Snowdrop' should be arriving with the first shipment, so I say give a hearty thanks to King Charles for securing New York for our combined benefit, an equally large thanks to his stuffy brother James for knowing when to quit, and the heartiest thanks of all to the fresh-faced new King William in the hope He will help us make the most of the present situation…"

Lurking within this smoke-filled melee, Jacob Wiley was operating as he always did, as a bottom-feeder in the depths below all this talk of business, creating a business of his own securing signatures for his insurance deals, offering rewards of exclusive tickets to the grand opening of Henry's upcoming Waterworks, in London's Piccadilly…

CHAPTER 41

OPENING NIGHT

"Roll up, roll up, read all about it – our Henry's latest attraction opens in Piccadilly with pretty damsels an' fire-breathing dragons in a show of magic the likes of which have never before seen by man," the newsboys chanted cheerily as if announcing the onset of Spring, whilst over their heads organ music blared, powered by a giant windmill atop the theatre which announced with a flowing banner - 'WINSTANLEY'S WONDERFUL WATERWORKS'.

Swarms of the Wealthy and Privileged descended like bees to honey, depositing their coins into the silk-lined tills with a singular aim - to mingle amongst the Great and Good hoping some of that 'Greatness' would rub off on them.

Henry meanwhile had hopes of his own - that the King would attend to bestow His Royal Approval – a hope shared by a couple of gushing socialites swarming nearby, whose eyes appeared to be under attack from exotic moths, their fluttering eyelashes being so outrageously large, "I do hope His Majesty will attend."

"It is unlikely – I am told He is leading a great War in Europe."

"I know so little about Him. What tickles His fancy then, for I have heard the fairer sex may not..."

"Aye, and I have heard King William will only blow kisses to a man."

"What does Queen Mary make of all this?"

"I have been reliably informed She is contending with her own personal battle."

"What could a Lady in Her position be possibly battling, without the support of her Army?"

"The Army can never help a Queen who is at war with Herself."

"How so?"

"The same source informed me that an enemy has invaded her skin, advancing steadily on all fronts, leaving pustules in its wake."

"Good Lord – another plague then?"

"Tell no-one, for the prevention of panic. It is our duty to put all selfish thoughts on hold and indulge in this frivolity, purely for the sake of its creator, Mr Winstanley."

"Agreed - I shall also play my part, for as a loyal subject I am compelled to do the honourable thing."

"Agreed, we must refrain from all frivolities, no matter how attractive they may be."

"Agreed. It is so refreshing to be in the company of a lady as noble as yourself."

"Agreed, for I am equally pleased to be in the company of a like-minded lady such as yourself."

"Do I spy Champagne in your glass?"

"What? Oh this? I was reliably informed it is purely water infused with bubbles."

"Good, in that case allow me to taste some of yours."

"Desist, Madam and find some of your own."

"You insult me, Madam. I shall depart forthwith."

"Agreed and good riddance"

The like-minded socialites scurried off to find the best available seats before the main official guests arrived – the businessmen, merchants and others of wealth hoping to meet the new King's cabinet, or at the very least, spot Tony Tyler of The London Gazette, who controlled the means of announcing their importance in the following day's edition.

With no pressure to purchase either tickets or merchandise, these wealthy freeloaders were about to witness a show the likes of which had never been seen before, whilst dining on the finest of foods illuminated by Henry's coloured fountains inspired by his visit to Versailles.

Whispers began to spread throughout the auditorium, leading to murmurs building to a thunderous applause like waves upon rocks once Henry arrived, (for even if few of the audience knew what he looked like beforehand, the abundance of self-portraits adorning the foyer ensured they all did now).

Climbing to the highest balcony, Henry gave the impression of a Lord on High keeping a watchful eye over all His subjects entering below, inviting the most influential to dine alongside him, being served on trays operated by ropes and pulleys. Tony Tyler was allotted a neighbouring table, providing Henry a clear preview of his report.

Within moments Henry noticed Jane arriving far below supported by Monsieur Ricard - looking suitably spectacular in his silver adornments - and raced down to join them, "Monsieur Ricard, it is so good to see you again, dear friend. I do hope you enjoy my show."

"I can see it is attracting a great deal of interest from your countrymen, an observation I will relay to my King."

"My thanks - there are delights aplenty on offer…" then, facing Jane, "Are you ready to be delighted too, my dear?"

"I am eager to see what has been distracting my husband from his husbandly duties."

"Then watch and enjoy, my love, for I truly believe to have finally found the reason I have been placed upon God's Earth."

Jane added with surprising sarcasm, "I do hope so, you have obsessed on the matter for far too long."

At this stage I should point out that it was unclear whether Henry had heard his wife's comment or not, for having been trapped in a bubble of self-satisfaction, Henry had become impervious to anything but flattery, more of which was about to be served along with a bitter after-taste:

"Your staging is undoubtedly impressive, my dear, as are the scantily clad damsels on display, however, my mother would have created more dignified costumes for such an occasion."

"Deflection is a major component of a Magician's arsenal, my dear, it diverts attention from all the mechanicals performing out of view, like a swan swimming serenely on the surface while its feet paddle frantically beneath."

"Indeed, then you are succeeding already in that regard – I can see gentlemen's eyes all about me, caught by the swaying hips, yet mine remain fixed upon those delightful railings. Do you not agree they are delightful, for you have never once mentioned them to me?"

Henry replied half-heartedly, still trapped inside his bubble of self-satisfaction, "Delightful as you say, my dear."

"They are the work of the truly brilliant craftsman I introduced to you."

"Indeed, my dear."

"Jean Fitou, sitting over there."

Henry may not have heard or being so distracted, failed to respond.

"I have agreed to meet Jean for tea close by, at the Lloyds Coffee House which I have been informed you are well acquainted with. Would you like to join us?"

But Henry's mind was already heading elsewhere and his feet soon followed, carrying him away.

Jane smiled at Ricard but spoke for her ears only, "Good."

While Henry and Jane were distracted in their own ambitions, three intriguing male guests had arrived in the theatre's foyer, moved into the shadows and chose to remain there - unexpected and undetected. The taller two men on either side were identical - dressed entirely in black, devoid of jewels and therefore of reflection or indeed detection, standing on either side of the third who was the most intriguing of all: a man of medium height draped in purple cloth covering both his head and shoulders.

Henry detected metallic flashes from within the capes of both taller men that, upon closer inspection, were emanating from highly polished swords of quality craftsmanship, devoid of insignia. Both men were as twins of robust and acrobatic stature with eyes constantly scanning the entrance, and as Henry watched transfixed, the purple cloth between them raised slightly, revealing an eye - staring straight into his, making him shudder as if being stabbed by a blade of ice, chilling him to the core.

Feeling suddenly vulnerable, Henry turned to attract his own security, but with shocking swiftness each of the taller men took hold of his arms - not enough to hurt, just more as a warning - the ensuing turning of their heads cementing the message.

As if that were not frightening enough, what happened next rooted Henry to the spot - the eye under the cloth winked at him. Henry stepped back, feeling stupid for having done so, but soon all was to be revealed: the purple cloth dropped to reveal a most elaborate wig, bouncing from its sudden release.

The man's face was familiar, yet presented more wrinkles than Henry remembered but the flash of His smile wound back the years in an instant. And then He spoke, clarifying everything:

"I remember you, young Henry - are you going to play any of your tricks on me this evening?"

"And I remember your wink, Your Majesty - yes, plenty of tricks, but alas, you flatter me - I am not so young anymore."

"Yet, still young at heart evidently," to which Henry performed an appropriately deep bow before the King continued, "Mary apologises for not being here – the duties of a Queen being so pressing of late; she is not enjoying the best of health. I am only here in an unofficial basis and glad to be so – nine years of war have left me yearning for both peace and amusement, so when I heard of your latest Wonder, I could not resist seeing what had become of the ambitious young man I met all those years ago."

"Ha – much older now, Your Majesty. Yet I am still ambitious enough to hope for your Royal approval?"

"Then may we start?"

Henry span on his heels alerting his Ringmaster to start the show. On the tap of his baton the huge, red curtains opened to release a flock of doves clattering around the arena bringing gasps from the audience, swiftly followed by two boys with flaming torches leading a scantily clad Galatea swooping low overhead, her golden wings surfing the wave of cheers.

"Ladies and Gentlemen, may we present, 'The cherubic Galatea's flight from the beast Polyphemus'."

Suddenly, a deafening boom reverberated as if all were seated in a giant drum, silencing everyone. As the seconds passed, Henry kept his eyes fixed upon Tyler and the Merchants, hushed in anticipation when a huge fireball exploded in mid-air making them gasp, announcing the arrival of the monstrous one-eyed Polyphemus hurling balls of fire in pursuit of his desire, followed by a trail of flaming dragons.

The audience burst into spontaneous applause so thunderous, it suited the epic tale well, bolstering Henry's popularity which he took to be assured when, glancing across at the neighbouring table, he found Tyler in excitable spirits, feverishly writing his review with head bobbing to left and right like a bouncing ball in an effort to avoid missing any of the spectacle on offer.

The show eventually reached its climax and as the curtains rejoined to form a red velvet wall, Henry leaned over eagerly, "Mr Tyler, would I be correct in anticipating a positive review?"

Tyler faced him with a glint in his eye that suggested he would, "You are to be congratulated, Henry - I have to confess I shall have to check the spelling of 'fantastical' and 'wondrous' for I am repeating them so often, it would ruin the Gazette's reputation if I should get them wrong."

"Can you offer a taste of what you propose?"

"If you insist, but the wording may change somewhat - '**Mr Winstanley's Waterworks is truly wondrous - a fantastical cocktail of hydraulic trickery, and pyrotechnics dedicated to the late King**'."

Henry received the praise in good humour, noticing the present King way down below leaving the theatre in good spirits with both guards in tow,

"My thanks, Mr Tyler, your words will attract even the most ardent of sceptics. I raise a glass of Champagne in your favour."

However, the liquid never reached his lips, for on the point of swallowing the first mouthful, Henry stalled - the most important guest of all was arriving far below, even more important than the King. Henry acted immediately, beckoning his Head Waiter, who was dressed as the Angel Gabriel, to guide this special guest upstairs. 'Angel Gabriel' dived backwards into mid-air eliciting gasps from the audience, deftly clasped a trapeze wire on the way down to land with the utmost grace at Sir Christopher Wren's feet.

As Henry leaned over the edge, watching this great man's ascent, he felt strangely nauseous - not from the height but the hopes and fears this most influential of men could bring, such that when Wren finally arrived extending his hand, Henry could only offer a weak response - his own hand shaking so,

"Mr Wren, I am so very delighted you found the time to come. I await your verdict 'with bated breath and whispering humbleness'*."

"Ah - a fan of the Bard no less! I ventured here in need of distraction, which you have provided most entertainingly. I thank you once again for stimulating my ulnar."

Sir Christopher laughed openly, Henry turned away, crestfallen, and the audience fell silent in anticipation of the main attraction which the ringmaster promptly announced,

"Ladies and Gentlemen, please welcome 'Mr Winstanley's Wonderful Barrel'."

A huge revolving Barrel appeared on stage, crowned by acrobatic nymphs pouring liquid into its funnel, which percolated to a tap below releasing beer, brandy, cider, wine, coffee, hot chocolate or tea on demand, as biscuits, cake and French bread were served dependent upon preference and budget. The final act brought the whole show to its conclusion, amongst a cacophony of fireworks as pyrotechnic flowers bloomed, and a robotic duck swam around a cistern, quacked, laid a metallic egg then waddled away.

The crowd were ecstatic, as was the applause which continued long after the show had ended…

In his hotel the following morning Henry awoke in good spirits, looked out over Hyde Park in good spirits, and remained in good spirits whilst opening The London Gazette. However, the good spirits ended once he turned to page 4…

"Only page 4 for heaven's sake!! Am I not a worthy addition to page 1, 2 or 3?!"

Henry took a deep breath, calmed himself and continued to read:

"Mr Winstanley's Waterworks is truly wondrous - a fantastical cocktail of hydraulic trickery and pyrotechnics dedicated to the late King…"

…so far so good, just as he had expected, but then:

'…alluding to Greek mythology, referencing the story of Galatea and Polyphemus which only serves to illustrate the superficial nature of all that surrounds it - not that it is tedious, for it made me laugh most heartily. I endorse Winstanley's show and urge you to see it, if only to improve the jollity of your day.'

The Gazette flew through the air with all the grace of an albatross and landed with the slap of a shot pheasant. Henry fell to his knees in rage, shredding the pages into pieces, each of which landed in the open fire, curling in the flames like slugs in salt.

Henry felt shock, he felt fury and he felt betrayal. The cognac helped as did the vodka, but Henry was sober enough to know these spirits were only temporary solutions.

The solution he needed above all others, was a black coffee amongst the honest company of straightforward men…

CHAPTER 42
DEVIOSITY

Dear Reader, at this point it would be worth reminding ourselves that, unless my writing has been unforgivably lacking, Jacob Wiley was clearly neither honest nor straightforward. By never hindering Henry's creativity (for why would he want to stop the Golden Goose from laying its eggs?) Henry had been oblivious of Wiley's true nature. However, more recently Henry had observed a degree of coolness in his wife's dealings with that man, that had led him to pay more attention - and in the paying of more attention he had noticed an essence of deviousness in Mr Jacob Wiley's character.

Therefore, it will come as no surprise that we find Wiley lurking beside Henry's table at the Lloyd's Coffee House on the fateful day Henry heard of a terrible loss...

The Coffee House was full of its usual clientèle that afternoon - a gentlemanly mix of middle-aged smokers somehow managing to suck at cigars, blow smoke, eat, belch and sip hot chocolate at the same time. Alongside a window we find a relaxed Henry dressed as expensively as his purse allowed - which had now grown to be more than adequate by anyone's measure - devouring a mountainous dish of pork sausage and mash, being observed with a degree of envy by Wiley seated alongside, picking at his pastrami.

"I fully understand your grief, Henry."

Henry looked up surprised, "You do?"

"Of course - after all you have both wealth and fame in abundance – what poor soul would not feel grief in your position."

Not being a lover of sarcasm, Henry offered a grunt before returning to the matter of mashed potato, forming waves with his fork.

Two portly Merchants arrived, ordered hot chocolate with cream buns and docked at a neighbouring table attacking the buns with a lust that explained their girth. The fatter of the two's jowls quivered as he began to wax lyrical about the newly opened Waterworks,

"Tell me, what was your opinion of Winstanley's show?"

Henry's ears pricked up.

"Pure whimsy, I tell you."

Henry groaned.

"I found 2 shillings and 6 pence frightfully good value for such jollity."

Henry slumped into his cushion.

"I laughed throughout the show and continued laughing, until I reached home and into my bed, on the way upsetting my poor wife so much she considered calling for the physician, believing my convulsions to be terminal."

"My heart was exhausted from all the laughing – oh, how my poor ribs ached!"

"Aye, my funny bone was fit to snap."

Far from soaking up the praise, Henry sank deeper into his chair. Wiley was perplexed, "Henry, what ails you so? You have reached the peak of prosperity - what can be so wrong with that?"

"Gadzooks, Wiley, I have been put on this Earth for a purpose beyond merely making money!"

Wiley was utterly perplexed, unable to offer sympathy - for how could any man place a higher value on anything other than money?

The more obese of the two Merchants obviously agreed with that sentiment,

"Henry Winstanley deserves to grow fat on the proceeds."

"Aye, aye, sir, yet all his wealth could not have saved those poor souls."

"And on the mouth of Christmas to boot. Was it not sixty lives lost?"

"Aye, sir it was - the whole crew. I am ashamed to speak of my relief at it not being one of my own ships."

"Pray, feel no shame, for in that respect I am as guilty as you."

"It had such a gentle name."

"'Raindrop', was it not?"

Henry's ears pricked up at that, too...

"That was not the name, but it was something remarkably similar –"

"Snowdrop?" Henry interjected, dreading the worst.

Both Merchants turned to face this stranger in their midst.

"Snowdrop? Aye – that was her name. Do you know her, sir?"

"I *own* her, sir!" Henry blew, "All sixty crew you say? And the complete cargo?"

Both men nodded in confirmation.

Henry stood abruptly and swayed, from the pressure of his blood and the flashes of his memory from his own shipwreck – 'The Lady Elaine's' face splitting wide open, the wreckers stabbing the crew, the jagged teeth of rock…

Henry grasped at his chair for support, creating a terrible din that put the Merchants into shock growling, "Hold fast, sirs - I have witnessed how a shipwreck can suck all that is good out of a man."

"We name the rocks 'The Devil's Teeth', sir for they relish chewing on the ribs of our ships."

Henry flew into a rage, knocking his salt cellar into the sea of mash, "Tell me, sirs - is there no guiding light?"

The duo shared their banter, "It is impossible, sir, no man could mount a light so far out to sea. It would surely take more than one man to achieve. I'll wager even Sir Christopher Wren could not rise to the challenge. I have word that he has not the mettle for it. Aye, I hear the fear of failure frightens even him."

Henry was shocked, but not unpleasantly so, "Really?"

"Put it this way, sir - if Wren cannot achieve it then nobody can…"

"Then I should go and see this foreboding place for myself."

"Do not be so foolish, sir - you would be entering the jaws of the Devil himself."

"This devil seems intent upon devouring my business and my men. If He also has a mind to destroy me, then Henry Winstanley Gent. needs to meet Him face to face, and quickly. Please excuse me, gentlemen."

And with that Henry made haste for the wash-room - not to wash but to get away from those obese Merchants and their infernal banter.

In the ensuing silence, Henry was able to enjoy complete peace, for no gentleman would disturb another whilst sitting upon that noblest of thrones, in that most private of chambers…

One and a half hours of perfect peace passed before Henry returned to his table, believing himself to be alone, but a shadow proved him wrong, sliding across the adjoining table making his blood freeze. As he stared at the tablecloth it lifted somewhat exposing an eye that swivelled to face him exactly as King William's had done on Opening Night, shaking Henry to his core. Whoever owned this eye shook Henry to the core again, cloaked as it was entirely in black making it appear to float like a snake's head bereft of a body. It did however carry a smell - the smoky scent of pastrami, leading Henry to conclude the obvious owner of that eye, which was soon proved correct - Jacob Wiley had been a silent witness throughout the talk of wrecks and drownings - for having no interest beyond matters of money, how on Earth could he be expected to hold opinions on such trivia as the death of a ship's crew? Having brooded for many months over a growing nest of insurance premiums, Wiley was unprepared for the questions that were about to fire his way from the most unlikely of people, in this most unlikely of places, starting with the more voluminous of the two Merchants,

"Sir, if you are the man who persuaded those bereaved families to pay for your support then God bless you. They will have need of your money now, as might we one day."

Wiley managed to produce a nod and a smile despite starting to panic - not helped by Henry's unwanted interjection:

"Mr Wiley shall move to cover all of your losses whenever they might occur, as guaranteed by his signature upon your policy and the listing on the ship's log. Am I right, Mr Wiley?"

Wiley blushed under the pressure of both Henry and the Merchants, made worse by Henry's following demand:

"Actually, I have suffered losses too – not only my entire crew but a valuable shipment of beverages intended for many destinations, including the establishment we are now seated within…" at which point the Manager arrived adding his own frown to the mix. Henry continued,

"It is with the greatest relief that I too signed the policy, and expect Mr Wiley to not only honour the cost of both my shipment and my ship, but to support the families who have lost their means of income through no fault of their own - is that not the truth, Jacob Wiley?"

Henry had never seen Wiley look so panic-stricken - having perfected a mask of indifference since childhood, "Henry, we need to talk - this will completely ruin me."

As Henry watched the man he never fully knew before now, begging for help he wondered why he felt no sympathy - could it be the fact that Wiley's interest in his Wonders was purely financial? Or the hawkish flash he'd seen in Wiley's eyes, when he first caught sight of him watching his lame wife's limp?

"Mr Wiley, I've always told my wife that I will never promise what I cannot deliver. If you cannot do the same, then I cannot help or continue employing you. Our values would have strayed too far."

To Wiley's horror, Henry slid back his chair and swiftly departed from the building, leaving him to face those plump Merchants with their costly claims - alone.

The more obese of the pair was the first to speak,

"So, Mr Wiley, at last it is good to add a face to that signature. I believe I include my colleague in wishing to confer our business in a more suitable location, preferably an office with a desk placed between us and the documents we both signed placed on top."

Wiley tried to swallow but failed,

"Why would you want to do that?"

Here I should ask you to consider Jacob Wiley's predicament - the horror of an advancing torso is directly proportional to its size and weight, so you can appreciate how horrified Wiley must have felt when the two portly Merchants closed in on him with thighs thudding and jowls wobbling.

No words were required - Wiley simply nodded his head, led them outside and beckoned a passing carriage wherein they all spoke very little, allowing Wiley to focus instead on formulating a plan...

The carriage headed East crossing cobbled streets, that did nothing to calm those wobbling jowls until Wiley finally spoke,

"My apologies, Gentlemen, but my office is in the midst of re-decorations so will not provide the privacy necessary to conduct our business. Might I suggest you both alight at the next corner where there is a very fine Teahouse. I shall bring all the documentation and monies to you there," - and as a final inducement:

"Pinkleton's are famed for their splendid cream cakes - 'a wonder to behold' I am told. Eat as much as your heart desires, for it will be my honour to pay for them."

To Wiley's relief he had touched upon their weakness - both men nodded their agreement and swiftly disembarked for the said Teahouse allowing Wiley's carriage to continue East into the land of Docks, soon coming to an abrupt halt as the Driver realised the destination.

"This is as far as I will go in such darkness, sir."

Wiley had no time to argue - the Driver's appraisal was justified - it was indeed very dark without the aid of moon or stars, and the lanes leading off bending in all directions were especially narrow like worms venturing blindly into a bricked maze of walls.

"Will you at least promise to wait here until my return - there will be a sizeable tip for you?"

The Driver nodded his consent, so Wiley handed him two silver coins, span on his heels and raced ahead into one of the lanes disappearing around a bend into the mist leaving only his footsteps behind, echoing off the walls.

The Driver, an honourable man, sat and waited as agreed despite the impending mist and freezing cold, thankful for the company of his trusty horses breaking the silence with their jangling harnesses, their grunts and their snorts.

Meanwhile some way ahead, Wiley was sprinting along the narrow alleyway, stopping every so often to check if he was being followed, then snaked his way towards an anonymous warehouse, when a man's gravel voice stopped him cold in his tracks. At first, he wasn't aware of anyone lurking in the shadows apart from that gravelly voice now clearing itself of gravel, but the thickening mist could not hide the movement that put the fear of God into Wiley. A shape was approaching, indistinct and getting closer until it finally appeared barely four feet away - an ancient one-eyed beggar thrusting a wrinkled hand through the bubbling mist, like a dagger. Wiley instinctively stepped back watching its fingers curl demanding a coin. But it was asking the wrong man, for Jacob Wiley had never given anything away in his life to anyone and was not about to start now, cursing the beggar before disappearing into the mist and heading for the warehouse.

At this juncture it is worth recounting an assertion that at the end of days our salvation may well come from duplication, for being constructed in pairs the body can compensate one organ's failure by employing its twin - and as a result a single kidney will grow larger to manage the work of two, as will a lung, and in the case of our beggar, his left eye had grown to work harder in order to compensate for the right. Thus, it was his left eye that clearly witnessed Wiley passing a coin into the Night Watchman's hand and the key he received in return.

Whilst shuffling along the maze of corridors inside the warehouse from one locked door to another, the only movements to greet Wiley were the swirls of dust rising from his own shoes and the occasional scurry of a mouse before he arrived at a familiar door, whose lock accepted his key revealing a room packed to the hilt with the wooden trunks he had painstakingly ordered into neat rows identified by handwritten tags.

Upon reaching the tag declaring - 'F.A.O. Mr J. Wiley of Wiley Associates' Jacob removed it and eased open the trunk.

His eyes sparkled as any father's would, upon discovering his baby safely cradled in a nest of comfort - in this case bulging bags of coins nestled amongst wads of banknotes.

Scooping as much as he could muster into four leather bags barely large enough to contain their bulk, Wiley headed for the door struggling to cope with the combined weight and gasping from the strain.

Meanwhile, 50 yards away in the swirling mist, the Driver had grown impatient to leave - a feeling shared by his horses with their scraping and clopping of hooves, forcing him to tug at the reins to settle them and in so doing missing the clack of the warehouse latch 75 yards away depositing Wiley into the freezing mist.

Being a professional man, the Driver had never abandoned a customer before and reconciled himself with the sizeable tip he'd received: but upon pushing the coins into his pocket, to his great surprise, they had bent then burst into powder. Incandescent with rage, he pulled on the reins such that when Wiley caught sight of him the carriage was already turning away and as Wiley screamed for the Driver to stop, he was already rattling off into the mist. The Driver had failed to hear the voice screaming from behind - or maybe he had chosen to ignore it, but either way Wiley was now alone – a desperate man out of breath and struggling with the weight of his entire wealth dragging from his shoulders.

Yet, as a disaster for one may prove a benefit to another, the one-eyed beggar was in luck - his good eye having spotted three Five Pound Notes fluttering down from one of Wiley's bags. Scarcely believing his luck, the beggar growled a gravelly growl and sprang forth with a vigour not experienced since his youth, chasing the trail of notes now falling from Wiley's bags like leaves from an Autumn tree disappearing into the mist.

To any casual passer-by 'Pinkleton's Tea-House' appeared to be closed - its lights had dimmed and there was no sign of life behind the fogged windows - but they would be wrong, for the two portly Merchants were still inside providing money for the tills and work for the staff. Their faces suddenly appeared at the window, chins covered in cream, eyes open wide, baffled by the sight of Wiley weighed down with leaking bags of notes, being pursued by a limping one-eyed beggar, with gravel in his voice and greed on his mind.

Never to miss an opportunity for wealth, the Merchants wobbled for the door and managed to squeeze through in time to witness the beggar making a grab for Wiley's leg before the horrific squeal of a petrified horse pierced the air, followed by a great thud and clatter announcing a carriage had smacked into both men bursting Wiley's bags before bursting him as they dragged the carriage over both motionless bodies.

Despite his horrific injuries, Wiley remained conscious for long enough to see opportunist thieves of all persuasions appear from the shadows, picking at his wealth like a flock of seagulls, stupidly tearing the notes apart whilst grabbing what they could, leaving the King's face soaking up the rain and rendering the Bank of England's 'Promise to Pay the Bearer on Demand' useless: - so much money wasted, and so many promises broken.

The last sense to leave a body is neither vision nor taste but sound, so as Wiley took leave of this world, he managed to hear the Merchants hailing him a hero for protecting their investments, and then nothing more…

Despite never having owned a sense of humour, Jacob Wiley could not suppress the final laugh which attracted the blood to his throat that made him choke on the sheer irony of it all.

If anybody had come across that grinning soul lying so freshly dead on the road that night, they could be forgiven for believing him to have left this world a satisfied man.

But nothing could be further from the truth.

Jacob Wiley left this world as poor as the day he had arrived, having learnt nothing in between.

CHAPTER 43
LIFE'S TRUE PURPOSE

With Jacob Wiley's passing it was as if Henry had become free of all distractions to his true purpose in life - yet, if only he knew what that true purpose was, he could then act upon it with total commitment...

He decided therefore to follow his gut, allowing his life experiences and good old common sense to show which path to take.

That path was called Water Lane, a place Henry had never visited before, but we have, as it leads to the door of Trinity House.

Thick fog was rolling off the River Thames stroking the doors of Trinity House - a place of noble intent like the elderly fellow manning its desk who we have met before - Harry Hull, a man who was afraid of water, now protecting those who had made it their livelihood.

"Can I be of assistance, sir? You look somewhat lost."

"'Tis a peasouper out there, I almost failed to find this place at all."

"I understand. You do not want to go the wrong way, sir, stepping into the river would never do."

"Indeed. I am Henry Winstanley come to speak with your Master and my friend Samuel Pepys."

"Mr Pepys is no longer with us."

Henry gasped, "He is dead?"

"Good gracious no, sir, but he is no longer the Master of Trinity House."

"I am glad to hear he still lives, but it is a great shame not to be dealing with him directly - being so abundant of wit and compassion, both of which would have served me well this day. Who is the new Master, and is he equally qualified?"

"We have Captain Mathew Andrews this year, sir, and as for his gifts I can vouch for the fact, he is a sailor of experience."

"Well, that is something, I suppose."

"And owner of the ship 'Barnardiston'."

"Ah, a ship's owner is even more promising."

"Permit me to lead you to the upper deck then, sir, but please be careful, the steps can be slippery."

Henry paused for a moment, not knowing how to respond - the comment was odd but had been delivered with such kindness, he felt it should it be treated with respect.

It took an age to climb the stairs as the old man swayed to left and right to such a degree Henry was prepared to catch him at any moment…

At this point it is worth commenting how easy it is to dismiss those who sit behind reception desks, for although they may not seem worth mentioning, they all have a history to tell - in the case of Harry Hull, it being a tragic loss a long time ago that had shaped his career and started his fear. Fifty years ago, whilst playing with his young brother at Tilbury Docks, Tommy had slipped into the river beside him. He did not shout, just a slight plop and he was gone, leaving Harry with a vision that still haunted him decades later - Tommy's mop sinking amongst the bubbled cloud released from it. Harry had stayed watching for any sign of hope and for a moment Tommy appeared to come alive way below, punching and kicking as if in a dance, but then his head kicked back, his eyes found Harry and remained fixed as he descended through layers of weed until the current took hold and whisked him away into oblivion.

Harry had been petrified of water ever since, had never learnt to swim and held those who could in the highest regard. Since that day Harry had felt duty bound to protect all men of the sea, and therefore had chosen Trinity House, a place pledged to the safety of sailors, as his calling.

Henry opened the Master's door onto two men staring back - a young one whom he instantly liked, and the other whom he instantly did not. Captain Andrews was like an old glum galleon having seen glory days on the High Seas, now resigned to forever remain docked on a river. Thankfully for Henry, the clerk still possessed the spark of youth. Both men were coated in the uniform of the place, with Captain Andrew's stripes providing the only dash of colour, made all the more striking for being set against such a plain and dark background. Henry's guard was up - for 'Uniform' sounded too much like 'Conform' for his liking, and not the sort of mind he had hoped to do battle with that day.

Using surprise as his weapon, Henry fired the first shot:

"Why in Heaven's name have you failed to erect a light on the Eddystone?!"

The effect was astonishing - the Captain's eyes flashing wide open, having not expected a salvo at such an early stage.

He growled loudly in an attempt to conceal his surprise, "Preposterous, there is no way any such thing could be built at sea. Who is asking for it?"

"It is not a request, Captain."

"So, it is a demand then. Such impertinence!"

"My name is Henry Winstanley."

"Never heard of you."

"I have recently lost goods and many good men on those damned rocks."

Captain Andrews eased back slightly, "That is the calculated risk all shipowners face, including myself."

"Then you should understand my plight, sir - I cannot just conjure a solution from thin air for something of which I know so little."

"Then you are of no use to me."

To the Captain's dismay, his clerk leaned forward enthusiastically,

"Mr Winstanley is not short of ideas, Captain, making quite a name for himself with a magical show on Piccadilly."

"I have no need of magicians either."

Henry leaned forward attempting to charm the Captain into submission,

"You are Master of the Sea and her ways, Captain, but they elude me as much as they eluded Alexander who could only achieve a light attached to the shore. I suggest a light built entirely at sea thus becoming no less than 'The Eighth Wonder of the World' - therefore Magic is exactly what I need, for no man could achieve such a thing without your help."

"Then forget the idea."

To the Captain's annoyance, his eager clerk leaned forward again,

"What about Sir Christopher Wren, sir?"

The Captain paused, lowering his guns, "Well, now, that would be an entirely different proposition."

So, Wren had entered the debate and Henry groaned at the inevitability of it all.

The clerk interjected one too many times, even annoying Henry with his lack of tact as he continued, "However, I have heard he is not too enthusiastic about the idea."

Grateful for the news, Captain Andrews stood bolt upright,

"Then neither am I. The matter is closed."

Henry's patience had worn away, "I never promise what I cannot deliver - as you refuse to do your duty, Captain, I am left with no alternative but to finance it myself. From the proceeds of Magic."

"You are a cheeky rascal. Well, go ahead and the best of British luck to you!"

Henry left without a bow, thudding down the stairs as loudly as he could, startling old Harry Hull behind his desk,

"Be careful, sir."

CHAPTER 44

A MEETING WITH THE DEVIL

The slap of hitting a new wave followed by the swish of sliding down another to greet the next developed into a rhythm that dulled the senses and allowed Henry's mind to drift.

He gazed across the constant sea hoping for a glimpse of the rock ahead but saw nothing, save for the occasional gull passing overhead, making him marvel how any sailor could possibly know where on Earth he was without a compass across this rolling blanket of nothingness.

Then he saw a strange but welcome sight - a shape, way ahead that rose and fell the same as him with its own slap and swish, and soon he made out the shape to be a boat, much smaller than his own, with two or three dots on board, and as he drew closer, the dots took the form of men and were four in number - the two facing him rocking forward and back whilst the other two rested.

"They are mad," exclaimed Henry to his middle-aged skipper - a man of few words wearing a leather skinned face blasted by salt, moulded from hours baking in the reflected sun.

"No, they are wise for they are heading to a rich fishing ground."

"But how on Earth can they know where it is, when it all looks exactly the same?"

"Because these are men of the sea, sir, and this is their home: the rolling sea is their carpet, the clouds are their roof and if you look to the horizon, that distant ship is the picture pinned to their wall."

Henry spotted the ship so tiny on the horizon, unable to tell its type or flag.

"We are here," the skipper roared.

Henry snapped out of his reverie - the skipper had sounded confident, yet Henry saw nothing new, "We are where, sir? I fail to see anything new." "Then look again. We are here."

Henry looked again but saw nothing new, until there - way ahead of them, he did see something new - the odd splashes of water rising as they met something solid beneath.

And so it was that gradually getting closer he saw more of the splashes to left and right, and they began to make sense, defining the edge of something huge below.

Fear took hold as Henry's boat continued bearing down on the thing -

Slap - Swish - Slap - Swish -

and then he saw it - magnificent in its simplicity - solid hard rock amongst so much froth, like teeth amongst saliva.

Closer still and its jagged edges became exposed by the swell, and at that point Henry understood where he was, and clearly saw the danger his crew must have seen too late before those teeth rose and fell, chewing the body of 'Snowdrop' allowing the sea to taste their blood before the teeth rose once again to crush their bones, setting into motion a gorging rhythm as if in a giant's mouth.

Henry had arrived at his crew's grave, and it frightened him to be at the mercy of such might, way beyond the frailty of this little boat which seemed to shrink the closer they got to the rocks, reminding him of man's frailty, making him call out,

"Please take heed, sir, any closer and we shall be crushed."

But the skipper seemed nonchalant, "Aye, they put the fear of God into me too, sir as they do on every trip. The reason I am still here is because I do take heed. I cannot predict them beyond the fact that if I do get it wrong - if I aim my boat too far to port or starboard, then I shall end up as the Kraken's supper - so yes, I shall certainly take heed, sir." And with that the skipper emitted a harsh laugh created from a humour most dark.

Henry could hardly believe his ears, "Kraken you say? Surely you do not believe such nonsense?"

Hoping for a smile Henry was shocked to receive nought but a scowl,

"Nonsense is an easy word for landlubbers whose views change as fast as quicksand, but we who hunt the sea do not change our minds so swiftly knowing how tables can turn in an instant. I cannot be sure what exactly goes on in the depths, sir, but as sure as I sit here, things will be hunting other things down there, and those other things might one day be us. I have heard tales of the Kraken in so many taverns that I will not say they are lies."

Harry laughed, "But a beast with ten arms?"

Yet, the skipper replied straight of face, "I have seen creatures with many arms pulled from the depths, and some as big as a boat with beaks that could snap your arm off. It is a strange and cruel world down there, sir - I am happy to live in the world I understand up here."

As he stared below in terror Henry was grateful, he had not yet felt the jolt of the rocks against their hull or heard them scrape against its side as his poor crew must have done on that fateful night, before being chewed and swallowed as surely as a fish on a plate succumbs to the jaws of men.

"I have seen enough!"

But the skipper just laughed and continued circling those teeth far too closely in Henry's opinion, as if taunting him and turning him apoplectic,

"I tell you I have seen enough, now turn back."

"I will decide if and when we turn, sir, you have no jurisdiction on this boat."

"Turn back at once, I tell you."

It was at that moment when Henry realised how the fate of all on board a vessel rests with he who commands it, and the consequences of his temper.

Powerless to make a difference Henry gave in, and slumped over the side, leaving one of the crew to comment: "Do not be ashamed of sea-sickness, sir - it affects different folk in different ways."

But it was no sickness of the sea that had gripped Henry - for staring into the depths Henry fancied he could see how his cargo must have sunk that day - the timbers spilling like fingers pointing to the depths, then shooting down to join it as trapped bubbles suddenly released, quivered and trembled their way back to the surface only to pop into nothingness.

With rising anger Henry looked again, and fancied he saw a vision more terrible than before – of his men following the cargo sinking beyond hope, trying to grab onto rocks for support ripping their softened skin in the process, before inevitably succumbing to Gravity tugging them down to the deepest depths to meet, who knows what – Poseidon or maybe even the Kraken itself...

Then, a shape appeared down below - huge and grey weaving its way between rock and kelp leading Henry to wonder - would this beast have been there that night - does it enjoy the taste of human flesh and if so, would it have feasted on his crew falling past its nose? ...

Henry remained staring for such a long time even the skipper decided enough was enough and turned his boat around.

Henry continued staring into the sea for the whole journey back - 9 miles of sea passing below his eyes turning into a jumble of shifting shapes meaningless to the mind allowing jumbled thoughts to enter unhindered, bringing with them a realisation, a determination and ultimately a conversation with the most important person left in his life residing 280 miles North, North East at their House of Wonders in Littlebury, seated in the wheelchair he had designed for her birthday…

"Have you any words for me?"

"I have returned."

"Evidently, but for how long this time?"

"Do not chide me, I am not a stranger."

"Then stop behaving like one," she growled sending her abacus spinning through the air smacking him on the cheek. Jane felt no remorse, "Take it – I had no need of it before, and I can manage without it now."

Henry continued walking straight towards her accepting more of his gifts returning as missiles - the bottle holder, the ice bucket, the cash till all denting his face, as she screamed issuing venom as only a tortured wife can,

"…Toys! Just toys! Your father was right! Nothing more than whimsy…"

Henry could not disagree with that, as in the garden a visitor was being tipped into the pond.

"…I need a man not a boy!"

She rammed the wheelchair into his shins bringing Henry to a bruising halt,

"I need you now more than ever, my love."

"Drop dead!"

Henry gave up the fight, dropping to his knees as he had done on the day he proposed 37 years ago.

"I had forgotten the colour of your eyes."

"I have never wasted my time on sycophants - your charm is like polish – it is wearing thin."

"Please listen to me, Jane…. I beg of you, please."

This time there was something in the tone of his voice that calmed his wife sufficiently to accept his affections…

With the wheelchair back in place, Henry pushed Jane in the chair he had built for her 20 years ago around the grounds of their House while his floating umbrella provided her cover from the sun.

To the fitting accompaniment of visitors' laughter, Henry bared his soul,

"A task has come to my attention that I cannot ignore."

"Another one of your 'Wonders'?"

"The most wonderful of all. Yet the most fearful."

"For Good or for Glory?"

"For God – and it frightens me. It is as if everything I have ever learnt, all my achievements thus far have been for this moment - from seeing the wind bend the trees on our ride to my shipwreck on a French beach - they have all come to this. It will take time, but I must do it. I have no choice. And you, dear Jane, are the only one to whom I wish to return."

Henry finished by giving his wife a long, lingering kiss cementing the gravity of his intent…

Over the following weeks, Jane allowed her husband the time he had asked for, observing him moving with an unusual calm around the grounds, stopping every now and then to sketch how the trees of all types bent as one to cope with the high winds, then disappearing into his glasshouse studio spending all night sketching despite the lashing of rain, the flashing of light and the bellowing of thunder.

As Jane lay in her bed throughout the following morning wrapped in a cocoon of comfort listening to the rain beating at her windows, her thoughts centred on this unusual man she had chosen to spend her life with - as opposed to *share* her life with - for she never fully understood the ways of his mind, or indeed why he would choose to spend a whole night in a glasshouse surrounded by branches beating at its windows…

But later that morning, she witnessed a surprising change - not only was the sun stroking her house with its rays, but the birds had found branches steady enough to sing upon, and when she dared peer out at the shattered studio, its door duly staggered open, and her husband emerged looking as if an anchor had been hoisted from his shoulders. He gazed up at the sun, and Jane saw him smile before hopping over to the house like a perky Robin Redbreast holding rolls of paper under its wing.

Unravelling his sketches onto the kitchen table, Henry chirped like a chick, tapping them here and there whilst pointing to significant things she could not understand the significance of, then hopped back outside like a wren even mentioning Mr Wren as he opened the gate to his garden of wonders - and for the first time she could recall, his mention of Wren bore no malice, just pure affection accompanied by a laugh.

Jane stepped towards the table curious but cautious, to see what her husband had been concocting and shook - the trunks of several species of tree she recognised from her garden were aligned in columns with notes attached comparing their proportions. Further sketches revealed how each had behaved in last night's brutal storm, blurred by watermarks from the dripping roof. She was intrigued and somewhat delighted to notice the addition of his trademark lantern, first employed above Saint Mary's Church in Saffron Walden, and presently above the entrance to their house - but it was his final sketch that took her breath away, for viewed as a whole her husband had in mind a massive and bold structure, the likes of which she had never seen before, and for the first time in his life - a thing of momentous intent.

Jane was left breathless, wanting dearly to speak with her husband, but had no idea where to find him, and even if she had, she knew he would be lost in his thoughts in a secret place.

On occasions such as these, Jane found it best to leave him alone.

In fact, if she had searched, Jane would have found him in the graveyard of the church where his father had enjoyed unequivocal stature as the saviour of the village, and where Henry, now aged 51, stooped to his knees in prayer before his father's grave.

"Now you have all the time in the world to consider what you have said to me, father. Words I can never forget."

Henry looked up and spared a moment to observe the stained-glass window which had enabled a clear view of his first eclipse, and so with renewed self-worth and a sense of destiny, Henry rose to his feet and then into his saddle to pull a cart, packed with equipment, over a long journey to the great task ahead – all in a manner of sober determination.

And more than a tinge of fear.

CHAPTER 45
PEPYS V. WREN

Trinity House was not a place of solace for Samuel Pepys having returned of late to the position of 'Master' in that sombre building, with its corridors populated by the 'Brethren of Old Sokers'*, as he called them, shuffling in and out of the shadows with minds full of importance whilst ferrying papers containing very little. With darkness of mood, this Great Man opened the door engraved 'Master of Trinity House' and stepped once more into the position that suited him so well, and the office he so deserved.

Sanctuary.

Yet, despite casting aside his dark attire, the clock above his desk ensured his mood remained the same, for being required to deal with the 'Old Sokers' every working day would test the patience of any man, even one as noble as Samuel Pepys. As Master of Trinity House, his position required both Tact and Diplomacy, and upon watching the clock's hand shudder towards 12, he knew that within the next 2 minutes he would be requiring both in abundance - for his guest was both strong of will and moreover, had a reputation for punctuality…

The knock upon the door alerted him, the swing of the door girded his loins such that at the time the clock struck the hour and his door opened wide, Pepys was clear of mind and ready to do business with his guest who, being a National Hero of the highest integrity, only made matters worse.

The guest, fully expecting the usual adoration, received nothing at all – Pepys being of the opinion that silence was the strongest tool in any negotiator's arsenal.

Unused to being ignored, the guest announced his presence with a polite cough.

Pepys's eyes finally raised to regard him, and so began the guest's soliloquy:

"This is grave news concerning our old friend Mr Eddystone, yet you will note it is exactly as I predicted: in making Plymouth the South West's major Naval base, it is inevitable more ships will be wrecked thereabouts – even if only from congestion. However, firstly I feel I should offer my apologies for failing to, to -" Sir Christopher Wren stammered, betraying his guilt, then to his great surprise, Samuel Pepys interrupted him, and quite sharply:

"'Constant' is the latest victim with all hands lost. Tragic news, Mr Wren, but ultimately fitting in a way…"

Wren took the bait and replied, including more of the aforementioned stammer,

"How so, may I ask?"

Pepys replied with a calmness that affirmed his authority,

"…for its famed owner shall now attempt to build the very thing I expect you came here to admit had defeated you."

Wren was at a complete loss - a rare event indeed, "Who?"

Pepys replied with surprising swiftness, unsettling Wren again, "He is known to his many admirers as the creator of 'Winstanley's Waterworks' in Piccadilly, which is still proving mightily popular with the public by all accounts."

Wren failed to suppress a laugh, "But this is madness - what on earth does Winstanley know of the sea?"

"Well, for a start his business relies on the sea, so he gives it the respect it deserves, and he knows it carries his goods from the Americas so will naturally want to ensure they arrive intact. As you know, we have fifteen lighthouses dotted around our coastline, yet not one of them was built entirely at sea…"

At this point Wren displayed a degree of arrogance at odds with his image,

"This needs an Architect not a mere Showman!"

The unflappable Pepys, however, remained unflappable,

"It might be that it needs someone with the bravado of a Showman, for no Architect has so far proved willing to risk their reputation on the enterprise – or their life," adding as Wren shifted uncomfortably, "I suspect Henry feels the fact no man in History has built such a lighthouse on a rock in open sea before, makes it a worthwhile challenge. The sweetener is that His Majesty, King Charles had desired it. This could well turn out to be Henry's Greatest Achievement."

"This could well turn out to be Winstanley's greatest downfall! Even to chip that hard rock sufficiently to form one small hole would take days, and that is if the weather is kind. Does Winstanley not realise he will need to create at least ten such holes to form a foundation!"

"Oh yes, but his plan is for twelve. The foundation will take many months to complete but he is aware of that. And he is prepared to row the 9 miles out and 9 miles back every single day to achieve it. Would any Architect be as willing to do the same?"

Wren looked down, unable to argue, allowing Pepys to continue his attack,

"And do you know what drives him to succeed in this task?"

Wren remained disinterested.

"It is you, sir."

Wren's eyes shot back up, unable to hide his surprise.

Pepys continued, "Henry told me his father was a great admirer of yours to such a degree he wished his son would emulate your achievements - a hard challenge for any Architect let alone a 'mere Showman'." Wren's eyes lowered once more in acceptance of his punishment, more of which was about to be delivered, "In fact you are both similar in one respect - you both prosper from the misfortune of others."

This proved too great an affront to go unchallenged,

"Really? How so, sir?"

Pepys continued unabated, confident of his convictions,

"Your structures rise from the embers of The Great Fire as shall Henry's from the graves of many a fine sailor. We have approved his design."

"May I see it?"

"I am treating it with the same confidentiality afforded to any design forwarded for our approval. You will understand, of course."

Wren was humbled, "Are there any provisions for his safety?"

"Do not fear – the Admiralty has despatched 'HMS Terrible' to guard Henry and his men from the main scourges of the sea – the inevitable pirates, and now also the French."

"I fear he may be sacrificing himself for nothing more than fame."

"For God - he says, for he already possesses fame. Yet, I say his greatest profit will be ensuring that sailors' families will not suffer pointless loss."

"But you surely agree this man is mad?"

"I would say Henry stretches boundaries, and as you should know better than most - History never remembers those who play safe."

"Has he considered the cost?"

"Yes - including the cost to life, as do I. The 'Old Sokers' upstream of us will no doubt be discussing this as we speak. They will consider the project doomed, and therefore unworthy of investment - whilst others of a more financial nature will point to the benefits of taxing all the ships aiming for Plymouth harbour. Then they will quarrel over the matter of costs and revenue, but ultimately, I predict Henry's plans will be approved. I try to ignore the 'Old Sokers' anyway," he said dismissing them with a wave of the hand, "Most are nothing more than corrupt doting rogues*. Let them devise other ways to build a profit than this. More important to me is the saving of sailor's lives from being guided by Henry's light."

Wren paused, Pepys' words having obviously struck a chord, "Winstanley should be remembered for this - providing of course, he succeeds."

"Absolutely - but 'should be' and 'shall be' are two very different things over which we have but limited control."

Wren sat very still for a very long time until, unable to hold Pepys' gaze any longer, propped up his collar to shield his face and swiftly left the building incognito, hoping none of his fans would greet him in the street.

Thus, one Great Man had left another Great Man, all the wiser for having met - and with God's good grace - somewhat humbler.

Much has been spoken, much written and much argued over many months to make possible what I am about to disclose to you now…

242 miles south of Trinity House lies Great Britain's great naval base of Plymouth. A further 9 miles away a barren clump of rock was about to kiss the boot of England's Greatest Showman…

Henry, now aged 52 and driven by determination, slapped his boot onto the red rock, and heaved himself upright onto the spot his mind had deliberated over for many of the previous months.

His small crew held back allowing him the privilege of an inaugural speech:

"By the planting of my foot -"

A wave suddenly crashed over him, hissing as it dragged him down amongst the swirl until one of the crew pulled him out with ease - his arms all the stronger for having rowed the 9 miles.

As soaked as a drowned rat, Henry attempted once more to assume the suitable posture required of a noble speech,

"Thank you, young Ben, by saving my life just now you are saving many more, for with our combined effort, this rock shall not claim one mo
 _"

But Henry was floored by another wave knocking him into the maelstrom below.

A young crew member came to Henry's rescue and spoke for the group,

"Very fine words, sir, but as we are the only ones listening, shall we at least start? There are but four hours of daylight left."

"Indeed, John. I just wish Tony Tyler were here to record it."

"I hear the wreckers are calling you a useless idiot, sir."

"Of course they are, for I am in danger of killing their trade. An idiot I may yet prove to be, many may call me a fool, but I am certainly not useless - or you would be the greater fools for working with me in the wet, freezing cold, eh John?"

"I just hope the French leave us alone, sir - My Mum says they would strip us bare an' cut our throats, for that is their way."

"I am led to believe the words 'cut' and 'throat' are more accurately applied to a pirate than a Frenchman, young John."

"I just hope we won't need to discover if that is true or not, sir."

"Fear not, John - as you can see over my shoulder - the Admiralty has placed 'HMS Terrible' to protect us. We are safe in the hands of the Royal Navy."

HMS Terrible had duly dropped anchor within view of Henry's efforts and over the following weeks its crew watched Henry's team unload boulders onto the Rock forming the foundation. 12 iron uprights were eventually planted into its base supporting a circular wooden hull inside which the boulders were placed - all very diligent, all very sensible, and all so very boring for young Captain Bridge, pacing his deck day after day checking for any sign of the enemy, secretly hoping for the Red, White and Blue ensign to fill his 'scope, for the young Captain held dreams of victory on the high seas, not watching 'A shed being built on a rock', as he was inclined to say.

Ian R Farr

But all that was about to change, and in his impatience Captain Bridge missed an historic chance of glory, by not remaining exactly where he was...

CHAPTER 46

THE BEGINNING OF THE END - COMPLETE

Under a baby blue sky devoid of cloud, Henry's men continued working, confident in the belief that the Admiralty was ensuring their safety.

"Oh, my Lord - look to the horizon, Henry."

"I always look to horizons, Tom."

Henry span on his heels following the line of Tom's finger pointing directly into the sun and gasped – an enemy Privateer under the French flag was slicing its way through the waves to reach him,

"I fear young Captain Bridge has left his post in pursuit of glory."

"Please remember to protect your throat, sir!" called young John, remembering his mother's advice.

"My thanks young John, now abandon rock, men - save yourselves."

But Henry's men chose to disobey him, standing resolute on that small chunk of rock guarding the minor miracle they had been creating under his guidance.

The French ship finally arrived, decanting twenty of its countrymen to negotiate the swell. Henry's sea-hardened men grinned as they watched the portly French Captain stumble to remain upright before collapsing to his knees losing the crispness of his uniform to the wash of the sea whilst adding several ribbons of kelp to his braids of gold.

The Captain's first word confirmed both his nationality and his mood - "Merde". Henry, however, vowed to maintain the manners instilled in him since birth as an Englishman of good character,

"I beg of you – stop, sir, you are wrecking my life's work."

"Your work here is over as soon will be your life. Debarrasez-vous d'eux" (Get rid of them).

The French sailors leaped into action stripping Henry's men of their clothes, petrifying young John,

"Be careful, sir, remember what my Mum said - it will be our necks next."

"You should always listen to motherly advice, John. However, in this case I believe she is mistaken."

But John's mother was proved correct – the French launched Henry's men into the sea to face a gruelling trip home with no clothes on their back, and no pay to supply the family table for their gruelling day's work.

The French Captain sneered, pulling Henry so close the garlic embedded on his teeth was overpowering. The sneer disappeared at precisely the same moment as Henry felt a blindfold being tugged across his eyes, 'Je m'appelle Capitaine Lazelle, Je suis heureux de faire votre connaisance, Monsieur Winstanley, surtout dans mes conditions' ("My name is Captain Lazelle, I am happy to make your acquaintance, especially on my terms.")

Henry struggled to reply through chattering teeth, "Pray, inform me of your intention, Captain."

Captain Lazelle answered in English, "I have been warned not to underestimate you."

"That is good advice, sir – but you have me at a disadvantage – who flatters me thus?"

The reply came as abruptly as anticipated, "You will discover soon enough."

Henry cried out – the cold rolling pressure around his waist signified a chain was being coiled around his body like a snake around its prey. Then Captain Lazelle spoke again, "Emmenez-le chez Le Roi"

Henry had heard those words before and they did not bode well, "Le Roi monsieur?"

"The King."

The explosion of laughter caught Henry by surprise, not from any French accent but the seabirds mocking him from above with their infernal cackling, as he was tugged blindly along a plank, heaved over a hatch and dumped to the bottom of the ship's hull. Despite himself Henry began to panic as something hot and hellish surged up his gullet forcing his jaw wide open to decorate the hull…

…four hours later Henry awoke to something nibbling at his toes in the belly of that ship, making a meal of the vomit he had provided. Companionship was a much welcome friend and if it had to be from one of the lowest forms of life then Henry was content for lying in his coffin at the end of days, he knew he would be completely at their mercy - the tiny creatures who cared not a hoot whether his allegiance be with England or France. In fact, Henry concluded that in many ways he was being treated fairer by these creatures than he was by the French.

His feet suddenly jumped from a force beneath the hull. Slow thuds beat against his soles leading Henry to conclude the boat was turning into the wind, hitting the waves head on.

Thud, thud, thud…

After a few minutes the thuds grew weaker, until Henry fancied he could feel the ship lift and glide across the waters beneath as if entering a place of shelter - his final port of call.

A voice in his ear, carrying the stench of garlic, grunted words he did not understand. Hands clasped his arms tight, and Henry felt himself being lifted to his feet then harried up the wooden steps from the bowels of that ship onto its deck.

The fresh air was a welcome change as Henry readied himself for the grand coach that would transport him to his judge and executioner – Le Roi - The Sun King. He was expecting a decorated coach at the very least...

The blindfold was tugged away but he saw no decorative coach, just a plain stone wall – a featureless ugly wall, and the mouth of Captain Lazelle contorting into a sneer once again, emitting the same vile stench as he spoke,

"Welcome to France."

Henry was pushed, tugged and prodded without the protection of cloth or wadding, from one stone wall to another, then down stone steps into a cell decorated with eight men chained to the walls like living portraits. Each body appeared to be terribly malnourished with necks as scrawny as featherless gulls barely strong enough to support their heads upon which faces were attached like masks of Death.

The Gaoler, a particular nasty looking individual wearing a goatee, a huge scar on his cheek and the instrument to create another, prodded Henry into position next to the final chain. Henry dropped to his knees guided by a new and brutal force - the Gaoler's boot applied to the back of his knees. All seven of the inmates watched his predicament without interest - for why should they care if an Englishman received the same punishment as them? However, the eighth prisoner was different, for his eyes sprang open swivelling like white balls in their sockets, until finding him at which point the pupils exploded large and black. To Henry's horror, the prisoner's lips formed a chapped kiss.

Henry called out in the little French he knew - "Gaoler, je demand - set me free - I demand - see - Sun King!"

The Gaoler lunged forward in an instant, pushing his knife point into Henry's throat eager to force it through to the spine.

"Vous voulez voir le Roi Soleil? Ha - Vous ne verrez meme plus le soleil!"

(So, you want to see the Sun King? Ha - you will not even get to see the sun again!)

All eight prisoners laughed as one, revealing many black gaps between their teeth.

The Gaoler took the candle with him, leaving Henry to cope by himself in the pitch darkness, consoling himself with the fact that at least the eighth man's chains would restrict him from being able to reach.

However, it gave small comfort - for he was so close Henry could hear the man's mucus rattling in his chest in need of a hearty cough...

A few minutes later, moonlight penetrated sufficiently for Henry to form an idea of what was surrounding him in that miserable place.

He saw himself laid out naked, vulnerable and not far enough away from those gleaming white eyeballs of the prisoner he feared the most. Thankfully, for the moment the prisoner was motionless, apart from the rise and fall of his chest as his lungs gurgled within.

And so it was that despite the chorus of coughs, gurgles and passings of wind, Henry finally succumbed to sleep.

Three hours later, Henry awoke with a jolt to see those same eyeballs staring straight back. He struggled to back away, but the chains would not allow it. Nervous and unable to flee, he hoped against hope the prisoner's chains would do likewise.

A strange but soft noise came from where the prisoner lay - 'fff-fff-fff

He saw an eye blink and the lips shaping kisses - only they were not kisses at all, for Henry had heard a deliberate sound the man was trying to make amongst the faintest of blowing -

'fff fff fff', eventually becoming recognisable as - 'fou fou fou'

And there he had it - a word he knew - 'fool'.

Henry smiled to himself, one must not take oneself too seriously, he thought. Maybe this man was right after all.

And so it was in a conciliatory mood that Henry finally fell to sleep - all night believing himself to be a fool...

...four hours later Henry joined the rest of the prisoners jumping to the sound of a whip-crack just outside the door. The door banged open, and the Gaoler was pushed back into the cell distressed and bloodied, begging for mercy by a force they couldn't see. Henry was aghast until it all became clear once a highly polished boot arrived with a thud carrying the man inside it - standing with straight back in the manner of the Military, topped with a tricorn. Henry understood why, when the second set of boots arrived immediately behind, bringing a familiar mouth into the cell - the sneer that belonged to Captain Lazelle.

"Take a piss now, Henri - it will warm your feet."

Lazelle laughed at his own joke - thereby revealing himself to be as weak a comic as he was a man.

As Henry was pulled to his feet and ushered out of the cell, for once, all the prisoners acknowledged him, either with jealousy or anger he could not tell, for when he glanced back at the door, their heads had dropped back to how they were before. Stepping along the same stone floor by which he had entered gave Henry some sort of consolation - at least he had survived the night unmolested.

Under normal circumstances, a long journey spent in conversation with even one's most ardent enemy should bring about some sort of mutual understanding, but this long journey to Versailles brought nothing - in main, because conversation requires listening as well as talking, and in this case, Captain Lazelle did all the talking, and Henry was left to do all the listening.

What he had to endure throughout that journey would drive most men to take their chances and leap from the carriage, leaving their chance to Fate, but Henry was unlike most men and well aware of his impending death; so, it mattered not how many times Lazelle repeated the impending execution, or his own glorious future, Henry could not find any phrase worth remembering as the French Captain continued,

"I expect to receive a medal at the very least for this - you are important enough a catch to warrant that."

Henry ignored him,

"...Oh, come on, Henri - smile while you are still able, for Madame Guillotine will soon put a stop to your face doing anything at all!" he laughed, gesturing gleefully at slitting his throat.

The carriage continued on its way transporting these two men sitting on opposite sides of the war to meet the most powerful man in Europe who was controlling it...

CHAPTER 47
REPETITION

And so, the stage was set for a melodrama of sorts between the two main protagonists Lazelle and Winstanley, seated on opposite sides of the War, whilst in the centre sat the most powerful man in Europe who was controlling it. Versailles Palace provided the backdrop, and two dozen Courtiers played the Support Artistes whilst the spotlight remained throughout on the one star of the show – the appropriately named Sun God Apollo, better known to his thousands of fans as King Louis XIV...

As Henry passed through the ornamental gardens, everything seemed exactly as before - even down to the approaching band appearing as a colourful toy army with its whistles and drums, yet this time there was a difference – a blinding light was floating at the front eclipsing them entirely. Such was its glare that Henry had to protect his eyes as did Lazelle declaring, "The Sun God approaches in all His magnificence!"

Then something spellbinding occurred beyond Henry's comprehension, that both intrigued and hypnotised him – for as the orb came closer it started to pulsate accompanying a chorus of angels. Henry was completely gripped until he noticed the legs beneath the orb pirouetting on their toes - and not feminine legs, but muscular calves sprouting hairs. As it got closer, the illusion gradually faded, superseded by a set of mirrors strapped to the body of a man, like some sort of knight's armour - a man who was pirouetting with balletic grace, whilst playing guitar in a show of stamina impressing Henry enormously, and as he peered into the light's glare, Henry fancied he recognised the Sun's face behind all the greasepaint, and having nothing to lose he posed a question which all others were too frightened to ask:

"So, are you Apollo today, Your Majesty, or pretending to be Julius Caesar perhaps?"

Lazelle stepped forward to shield his King, cursing Henry's impertinence loudly enough to be heard by his King. Unimpressed and more than capable of defending Himself, Louis just brushed Lazelle aside as if he were a flea, and addressed Henry directly and quite calmly, commenting upon his nakedness,

"Je n'ai jamais vu ton penis auparavant, mais le reste d'entre vous a l'air familier"

(I may have never seen your manhood before, but the rest of you looks familiar.)

Lazelle took delight in translating, but Henry replied without fear, for what more could he lose when he was due to lose his head? "Whilst you look exactly the same as before."

Lazelle protested, but The King remained calm, presumably because nobody had ever dared to address him so boldly before.

"Vraiment? Comment ca?" (Really? How do you mean?)

Henry signalled to Lazelle,

"What did he say?"

"He asked in what way."

Henry chose to address the King directly,

"You were a younger clown then, Your Majesty, yet a clown nonetheless."

"Vraiment? Comment est-ce que j'ai marche? Comment ai-je parle?"

(Really? How did I walk? How did I speak?)

At this point, Henry noticed three pinpricks of light reflecting off three sword hilts ahead, peeping out from the capes of three armed guards posing as courtiers, so adjusted his manner accordingly, "You were but six years older than me, Sire, yet even then there was something special about you."

The Midget translated for his King, who seemed intrigued, "Continuer…"

"You may no longer be a clown, Sire, but you are still acting like one."

The Midget translated, and the King paused, and for a while Henry thought he would be executed there and then, but His Majesty seemed indifferent,

"Mais, je suis toujours le Roi Soleil." (Yet, I am still the Sun King.)

Henry understood well and pushing his luck still further stepped forward, using his hands to illustrate, "The Sun is but one star amongst many, Your High and Mightiness."

The Midget duly relayed Henry's message, albeit nervously, for he was obliged to perform his duty no matter the risk to his own life. King Louis paused, moving only his lips replaying Henry's words, so unused was He at receiving such a direct insult. After seconds that seemed like hours while Henry waited for the inevitable, and the guards clasped their swords, King Louis suddenly laughed heartily throwing his arms wide, and began spinning on the spot, "Alors, Je suis l'univers!" (In that case, I am the universe.)

The King span away playing his guitar like a spinning top, and, as in the Pied Piper's tale, everyone duly followed - the courtiers all mocking the size of Henry's manhood as they passed.

Henry guessed it may have been due to his age, but Versailles no longer held the magic he remembered, as if the cracks were starting to show - the huge room of mirrored columns being replaced by ridiculous derivations of one subject - Louis as a Ballet Dancer, Louis as Julius Caeser and even Louis as Apollo, the most beautiful of all the Gods.

As he continued walking, Henry failed to hide a smirk at passing a trio of courtiers kneeling in a corner relieving themselves without shame, bringing him to a wry conclusion: - he may not have entered the gates of heaven that day, but by nightfall he surely would.

The King strode past eyeing Henry's appendage, with his Midget in tow ready to translate, "Avez-yous engendre beaucoup d'enfants?" (Have you sired many children?)

"Unfortunately, not, Sire."

"C'est ce que je pensais." (That is not surprising.)

The Midget released a laugh and as was their wont, the courtiers followed suit, and they remained laughing, as Henry was led to the King's bedchamber, feeling ashamed for having to represent England at half mast, so to speak.

The King turned to address His courtiers whilst pointing at Henry,

"Qui a fait entrer l'Ennemi dans ma maison?" (Who led this enemy into my house?)

Captain Lazelle sprang forward with medals jingling on a chest puffed like a pigeon on heat, "C'est moi, Votre Majeste, Lazelle, Capitaine de votre Marine Royale."

(It is I, Your Majesty, Lazelle, Captain of your Royal Navy.)

Henry waited in frustration, impatient to get on with his sentencing, but Lazelle was on a roll: "Je l'ai trouver sur un rocheravec vingt matelots ennemis, tousce qui ont personellemont rejete a la mer comme un panier de poissons.

(I found him on a rock with thirty enemy sailors, all of whom I personally threw back to the sea like a basket of fish.)

King Louis rose to his feet, "Enlevez vos pantalons." (Remove your pantaloons.)

More used to giving orders than receiving them, Captain Lazelle faltered, as did Henry and the courtiers, who all became excited at the prospect. As Lazelle finished removing his breeches, and the King finished inspecting him, it was Lazelle's turn to become the victim, as his King ranted, "Je ne m'attends pas a ce qu'un Capitaine de ma Marine d'etre si inapte!"

(I do not expect a Captain of my Navy to be so unfit!)

At this point King Louis kicked his own trousers aside, impressing Henry with the robustness of his physique.

"Je choisis judiceusement mes Capitaines - ils representent la France et donc moi!" (I choose my Captains wisely - they represent France and so represent me!)

Lazelle stared in disbelief as his King continued,

"Les dirigeants d'hommes qu'ils dirigent et recevoir de l'amour en retour."

(Leaders of men should love the men they lead and receive equal love in return.)

Then continuing with a wave of his hand,

"Apportez la boite" (Bring me the box) at which point two men of low rank swiftly disappeared, returning moments later carrying a 6-foot simple wooden box between them as Henry looked on grimly, adding an appropriately grim request, "I beg of you, please let my final journey be taken in a more comfortable vessel than this."

The Midget translated for the King who burst out laughing, making Henry feel even worse.

He was led to a large room where courtiers were gathered around a long table with the King taking His seat at its head alongside which the Midget was struggling to jump into his chair. Henry stepped forward to help, for this frail little fellow had not weathered as well as his King - yet he still managed to bat Henry aside whilst maintaining an air of superiority. The King suddenly clapped his hands and spoke as loud as any King would, to command attention, "Prenez garde" (Take heed.)

The courtiers all stopped and took heed.

"Je vais dire quelque chose d'important." (I am about to say something important.)

The courtiers all raised their quills.

"Nous sommes en guerre contre l'Angleterre" (We are at war with England), to which all roared their approval, "Mais nous ne sommes pas en guerre contre l'humanite" (But we are not at war with humanity) to which they all fell silent. "Toutes les navires beneficieront de ce que cet homme, y compris de la France." (All ships will benefit from what this man is doing, including those of France.)

King Louis beckoned Lazelle forward, exchanging His countenance from one of wisdom to fury, "Je choisis judiceusement mes Capitaines, car les meneurs d'hommes doivent aimer les hommes qu'ils dirigent comme ils l'ont toujours fait, et recevoir de l'amour en retour..." (I choose my Captains wisely, for leaders of men should love the men they lead, and receive the same degree of love in return...)

Captain Lazelle was lapping this all up, for as his King was a God, His words were sacrosanct,

"Capitaine Lazelle la boucle est bouclee - a partir de ce jour, ceux que vous avez conduits vous conduiront maintenant. J'espere qu'ls vous donneront le respect que vous meritez. Donne-moi votre epaulettes."

(Captain Lazelle, for you the decks have turned such that from this day forth those you have led shall now lead you. I hope they give you the respect you deserve. Hand me your epaulettes.)

Lazelle was understandably shocked, "Votre Altesse, j'ai combattu de nombreuses annees pour l'honneur de mon Roi." (But, Your Highness, I have fought many years in your honour.)

"Pourtant, ces derniers jours vous m'avez deshonore, moi et la France! J'attends de mes officiers qu'ils se comportent commes des gentlemen. Je vais dire les choses telles qu'elles sont - vous etes le capitaine d'un yacht dans La Manche Lazelle, pas d'un gallon en haute mer."

(And yet, in the last few days, you have dishonoured both myself and France! I expect my officers to behave as gentlemen. Listen well, for I shall state things as they are - you are in charge of a privateer in the English Channel, Lazelle, not a galleon on the High Seas.)

"Mais je t'aime Empereur..." (But I love you, my Emperor...)

"Va-t'en, etudie l'humilite, et mettez-vous en forme. Executer maintenant!"

(Begone, study humility, and get yourself in shape. Now!)

For a few seconds, Lazelle remained stationary in shock, until the King span around and strode off, leaving Lazelle completely alone - a broken man, like a boat without a mast. Watching him crumble brought Henry no joy, for any man of position knows how quickly he can fall once his post has been removed.

A tap on Henry's knee announced The Midget had something to say, and if his expression was any indication, it was going to be sarcastic, "Once again, the King finds your appearance amusing."

But Henry was in no mood for joking and rounded on the little man with a finger thrust at his own head, "It is not how I am dressed that is important here, my diminutive friend, but what lies in here!"

Never having been addressed in such a manner, the Midget frowned for a moment before blurting, "So, tell me why I should not introduce your neck to Madame Guillotine?"

Determined not to be intimidated, Henry dropped to his knees and addressed the Midget face-to-face, startling the poor fellow, "Be careful, my little friend - upon this neck is perched a head full of wonders that may be of use to your King. Do not lose my head or they will be lost to your King forever."

But the Midget just huffed, "The King says your clothes still amuse Him - is walking naked the current fashion in England?"

"Tell your King my skin was provided by the King of *all* Kings. However, if Apollo is feeling generous, then a set of clean clothes with an extra-large collar to help guide your axe would be gratefully received."

"You will need all the help you can get," the Midget replied with a huff.

There was a question Henry was eager to ask, "Incidentally, where is my friend Omar Ricard?"

The Midget smiled, and for once not unkindly, for his manner seemed to have softened since Henry helped him onto the chair,

"At The Hague, eating pancakes with the Dutch – M. Ricard is acting as the King's Counsel."

"Omar is well suited to that."

"He will need to be - we are relying on His grandson to act like a man when he is but a child of six."

"Aye, it seems the fashion to be led by children, your King was barely nine."

"After four stillbirths, my King arrived as a gift from God - 'The God Given' they called Him and now He is the most powerful man on Earth. Follow me and I will prove it."

Henry entered the Royal Bedchamber passing two physicians testing the foul-smelling contents of a silver bowl – "His last supper?" Henry asked risking a joke.

"His last movement," replied the Midget, accepting the ruse, "He is an exceptional leader and an extraordinary lover, as you are about to witness."

A cheer rang out from the courtiers surrounding the bed who broke into applause once the King's Mistress, Julie de Chateaubriant, was led in by two of her maids. She made a small bow of acknowledgement to the audience, before the maids set about preparing her for the King's pleasure, loosening her clothing and securing her limbs to the bedposts. Henry's arousal was plain for all to see, as he struggled to cover his manhood, but it was too late, for Mistress Julie had already noticed, bringing a smile to her lips. Henry turned to leave but was beckoned back to watch, for this was a regular performance, and as she lay supine the remnants of her silks were pulled away leaving only her jewellery in place, thus enhancing her nakedness.

King Louis arrived as the star of the show, readily erect and grinning at the audience accepting accolades from all His spectators, bar one - Henry being too English to comply. Once the maids had performed the parting of Mistress Julie's legs, His Highness dived on top as if mounting a horse and set about humping in a manner akin to a stallion with a mare.

And so, the Royal 'performance' continued - for this could not be called lovemaking by any stretch of the imagination - His Majesty performing more like a beast for the benefit of His courtiers who clapped their hands in rhythm to His thrusts. Feeling intimidated, Henry made to leave, but Mistress Julie was insistent he remain, pulling his hand into hers, to which he awkwardly complied seemingly boosting the King's fervour, and bringing the show to a satisfying climax.

The performance now over, three male servants performed what appeared to be a regular ritual - the dressing of their King in riding gear. Louis roared like the King of the Jungle, then rode off at speed leaving his Mistress and Henry somewhat exhausted.

As Henry waited, unsure whether he was meant to follow or not, the matter was resolved with a smack to the buttocks and he was thrown onto a saddle to follow the King galloping through forest and over field pursuing a distant deer unsportingly restrained by straps held by servants acting upon the orders of their King. Once released the deer leaped in the air enjoying mere seconds of freedom before King Louis's pike impaled it mid-jump, and with a great cheer from the servants, the poor beast thudded back down to earth. This daily ritual continued as local farmhands raced forward producing blades to stab and slice, streaking the white birch trees red, soaking the brown earth maroon, and supplying the butcher with the means to make a meal fit for a King.

Henry was exhausted, and in a certain degree of awe as to how this Frenchman, some years older than himself, could conjure the stamina to perform such feats - a physicality he wrongly believed to be beyond his own capabilities.

Returning to the Palace in agony with thighs covered in lesions, Henry felt surprisingly elated from having spent the most bizarre day of his life with this greatest of men, who was living life fuller than any man he had ever known.

All of Henry's elation deflated the moment he stepped back into the huge pink room and saw the depressing reminder of his predicament - the wooden chest deposited on the floor ahead - his transport from this world to the next.

Nothing focusses the mind more than knowing you are seconds from death, so having nothing to lose, Henry demanded a last request,

"Your Majesty has me at Your mercy. May I once again request a set of clean clothes with an extra wide collar to aid your axeman?"

To his embarrassment, both the Midget and his King burst into laughter,

"Sortez cet Anglais de sa misere." (Put this Englishman out of his misery.)

The Midget gleefully repeated the demand in English for Henry to understand - which he did, bracing himself for whatever form of beheading King Louis had in mind.

But to his surprise, the Midget handed him Lazelle's discarded pantaloons and so dressed Henry was led across the finest of pink Persian rugs in more comfort than he had been granted of late, reaching the wooden box and the answer to his destiny. Henry watched in awe as the lid was opened - his fate having been sealed for the King's pleasure…

However, what greeted Henry's eyes was a thing of dreams - glorious dreams - for what lay before him was a chest full to the brim with treasures - a gemmaphile's delight.

"I do not understand, Your Majesty."

The Midget stepped forward and relayed his message with a degree of warmth hitherto absent,

"My King has in His possession a report written in M Omar Ricard's hand, of your accomplishments in England. Your House of Wonders particularly intrigues His Majesty. Would you consider creating one for Him here, at Versailles? The degree of His gratitude is spread here at your feet - use it to fund the completion of your Light, to the benefit of sailors from all nations, then return to build a house of Wonders for the benefit of my King."

Henry was at once surprised, relieved and humbled,

"Please relay this message to your King - I could wish for no greater honour. I shall make it my life's duty to create a Palace of Wonders for

His Majesty, the like of which has never been seen anywhere across the expanse of Europe, or indeed the Earth..."

Upon hearing his Midget's translation, King Louis was overjoyed, but Henry had more to add:

"...but only upon the completion of my work."

At that point, King Louis was handed an official letter.

"Donnez-moi un peu de temps pour reflechir a ces..." (Permit me a moment to ponder this...)

King Louis duly pondered for quite some time, such was the importance of the letter from Mr Samuel Pepys, then with a grand gesture He beckoned a servant who raced into position bringing a desk, whilst another brought the ink and waited with quill poised to immortalise The King's words on paper, which He delivered with appropriate aplomb declaring,

"Your honourable, Mr Pepys, I am at war with England, not Humanity. Mr Winstanley's Light shall guide my French Ships to a safe haven, as it will those of all other Nations, without prejudice."

Whereupon He beckoned the Midget to relay a Royal decree,

"My King says your Admiralty are pleading for clemency...very well - go forth and complete your grand endeavour, then return to create His."

Henry bowed as deep as his tight-fitting clothes would allow, then span on his heels and sped away, the sound of ripping fabric bringing a smile to all of those present, compelling the King to address them all in stern conclusion,

"Ne riez pas, Cet homme va ici pour creer l'Histoire".

(Laugh not - this man is leaving this place to create History.)

CHAPTER 48

THE COURAGE OF IGNORANCE

"Wish me luck."

"You are a fool."

"I am about to create History."

"You are deluded. "

"I shall row 18 miles each day -"

"It will be the death of you."

"Thank you for your concern, but I am robust."

"53 is not 23 - you will not see 54."

"King Louis rides his mistress every day, then goes hunting, and rides home to ride his wife again - and He's 59!"

"You cannot manage but once a month! Besides, He's French."

"Is that all you have to say before I go?"

"Oh, I have much more to say but there is no use - when you set your mind to something you are as stubborn as Ben Drury's mules!"

"You will hear the front door open in the morning as the clock strikes 3. I shall close it upon my return six days hence."

"Ha! If I multiply that ten-fold and add another ten months, then I might be halfway to the truth!"

"Is that it then?"

"That is it."

"Goodbye, then."

"Be gone."

She heard him climbing the stairs so stepped into the hall and added, "Please be careful," said in such a low voice he might not have heard, which suited her well because she did not wish him to.

And so, at the age of 53, Henry set off on the most ambitious endeavour of his life, one that even a man half his age would have found challenging. But something inside him knew this moment in time was special and as such required his full attention - indeed more attention that he had given anything of late, even the woman back at home who he had always devoted half of his mind to…

At 6 a.m. many weeks later, that woman was lying in a warm bed with nothing on her mind beyond the style of egg to have for breakfast, while at exactly the same time 281 miles South West of her, and six feet underwater, a Corkwing Wrasse was also starting its day, meandering amongst the familiar pull and push of the currents washing the Eddystone Rocks following the feeding trail of a much larger fish too distant to be of concern.

The Corkwing Wrasse is a magical fish whose size may render it of little significance in a fisherman's net yet, is proof of an astonishing trick of which Henry would be proud, for as one of the Earth's great survivors it will switch gender to preserve its species. Despite appearing a humble soul, it is no shrinking violet - like a showman wearing its patterned waistcoat of orange, pink, green or blue lit from within, illuminating a path between the rocks of its familiar hunting ground.

Lately the Wrasse had experienced some unfamiliar intruders - one day it was a nail glowing red with heat that had bubbled and hissed its way to the depths, on another it was a huge creature with arms and legs, releasing bubbles as it scrambled to return to its own world, but the next tiny intruder was the strangest of all - a hook flashing in the sun as it twisted and turned. And what is more, this most curious of things seemed to be delivering a gift - the smell of crab was compelling enough, but to see it dancing so close to its nose was too tempting to resist. However, temptation is a dangerous thing and once inside the Corkwing's mouth, things happened that had never happened to it before - the strange force that lifted it past familiar strands of kelp, which seemed to wave goodbye as it continued up and through the skin of the sea into a surprising new world it had never felt before - rising past strips of wood with metal fixings until being dragged through a square hole into another world of intense heat inhabited by another two of the strange creatures - the older one with kelp across his chin kindly removed the hook from its mouth, before slicing open its chest and removing slippery things it never knew it had owned before, being laid to rest on a sizzling circle of metal that dissolved all its cares away...

Returning to port for the last time, Henry looked back at the Rock a proud man, for after two years of toil his creation was complete. He had not only achieved a gift to mankind but an advertisement of his skills like none other: from the lantern on top so reminiscent of the Clock Tower at Saffron Walden, to the whimsical additions he had kindly added for the benefit of the keeper, sprouting from its trunk like the branches of a tree - a seat from which to gain access, a rod with which to fish, and a crane to lift his provisions.

Advertisements on all sides of the tower promoted Henry and God in equal measure, ensuring each passing ship would spread Henry's name from Europe to the Americas on the lips of passing sailors.

CHAPTER 49

BY ROYAL COMMAND

Tony Tyler was a pragmatic individual enjoying success due in large part to that philosophy, and despite being pressed for time, he had no wish to offend this friend of Henry's who had come all the way from Littlebury to meet him.

Jack's face looked familiar to him, but its carefree expression did not indicate a mind worth tapping. The result was a situation where Jack's persistence grew in direct proportion to Tyler's irritation, so as Jack leaned forward, Tyler leaned back maintaining a balance between the two…

"Have you lost interest in Henry then?" Jack asked, as loyal a friend as ever.

"Times are changing, as are the interests of my readers. We must remain flexible, or we shall perish."

"Some writers disagree with you - I hear that diarist Celia Fiennes rode side-saddle through Plymouth hearing news that Henry had recently lit all 60 candles on his lighthouse. He is a local hero."

"Henry has done very well, but he has done very well *in Plymouth*, not in London."

"Excuse me, Mr Tyler, but Henry's Waterworks still proves very popular here in Piccadilly - I saw crowds milling around it just now as I passed."

"Look, I appreciate your loyalty to your friend, but you are misguided - Henry is no longer newsworthy in London. In May I headlined 'THE PROVOKED WIFE'S' PREMIER IN LONDON'S WEST END' and days later in North London, 'THE WORST HAILSTORMS EVER SEEN' -

both stories proving mightily popular and both stories concerning *London*. Tell me a tale of storms or London, for they will keep me in business for weeks."

"I do have a story set in London if you are interested. It involves Royalty."

Tony Tyler suddenly turned on a pin completely captivated, revealing the depth of his integrity, "Royalty? Now, that may be of interest."

"I thought it may."

"When you say Royalty, just how Royal are you inferring?"

"I would say, the most Royal of all, sir."

"So - the King?"

Jack responded, playing the game but gaining no pleasure from it,

"I have taken too much of your time already, Mr Tyler, I will go and ask elsewhere."

Before he reached the door, Jack heard the voice he was expecting, saying the things he expected it to say, and found the whole thing depressingly predictable:

"I meant to ask - how is Henry these days? Come and sit, John . Would you like a hot chocolate, or a coffee perhaps - we could go to a Coffee House if you so wish?"

But Jack wished not, and left.

Dear Reader, Time plays its mischief with us all, curling our spine, greying our hair and thinning our skin into a costume styled for Death.

Yet, it affects some more than others, which, besides the natural rotting of flesh and bone, depends upon the attitude of the mind itself to Life and which explains Henry's smile that day...

Henry was smiling as he stepped from the carriage onto a fresh carpet of snow in front of St James's Palace as huge flakes settled about his shoulders like a Lord Mayor's necklace. He remained smiling as he crunched a path through the snow, passing the troop of mounted guards, with their snorting steeds and jangling harnesses, and into a room of complete, silent splendour.

He was grateful for the outfit Jane's father had made for the occasion ensuring he looked suitably grand for a meeting with the King, presented as he was before His Majesty's guests dressed in embroidered glory, looking as much a dandy as the King himself who was now approaching, pausing to regard himself in each and every mirror along the way,

"Pray, tell me where is your wife, Henry, if you have one, which for reasons of my own, I am hoping you do not?"

"I, indeed, have a wife, Your Majesty, but she has decided to remain at my House of Wonders, which she is overseeing as we speak."

"Indeed, or is she just tired of being around people praising her 'Most Wonderful of Husbands'?"

Henry shocked himself with a blush, for once not knowing what to say like a starstruck youth. Thus, as he knelt and kissed the Royal ring that day, though his lips may well have been chapped from blasts of salty surf, at 56, Henry's eyes still contained the youthful twinkle that King William as a Prince had remarked upon, 29 years ago at Audley End.

However, Henry's knees were now showing their age, for when King William beckoned him to rise, he achieved that Royal command with some difficulty,

"Forgive me, Your Majesty, but the Rock is taking its toll."

Time had been more forgiving to King William however, which was understandable considering His Highness was more used to navigating sheets of silk than Force 8 gales,

"I remember my Uncle predicting great things of you, Henry, and you have delivered upon that promise admirably. I had actually called for a lighthouse to be built on that rock years ago - I was ignored by those who had the authority to implement it, and now it has needed a resourceful subject like yourself to finally deliver upon my demand."

"Your Majesty flatters me too much."

"Not at all - my wife and I would like to salute your latest achievement in a style befitting the many sailors whose lives shall be saved by your 'Beacon of Light'. For each and every day it provides safe passage for those in peril upon the sea, you shall hold a special place in History and a welcome at my door."

The guests applauded politely, pleasing Henry no end, despite his elderly father-in-law snoring throughout the proceedings.

"Now it just rests upon me to ask all of you to raise a glass in honour of the extraordinary Mr Henry Winstanley - Rum being the appropriate liquor."

The room full of senior Naval figures, including Samuel Pepys, raised their glasses in Henry's honour who bowed deeply in return, delighted that at last the Admiralty was acknowledging his worth.

The King continued in buoyant mood,

"I invite you all to a party here at my Palace. The theme is Pirates, so I shall expect to see many an eye mask, much swordplay and a modicum of Jolly Rogers!"

Mr Taylor had slept throughout the proceedings yet upon the King's request for costumes, suddenly sprang to his feet, cried 'God Save The King' and grabbed his quill to take orders.

So, satisfaction had finally landed at Henry's door, and he found solace in the fact that as long as his Lighthouse stood on that rock, his Fame would endure.

To cement that fact, he commissioned a miniature version to be installed at the front of his House of Wonders to remind every visitor of his worth.

CHAPTER 50
THE GODS ARE ANGRY

Beyond the door we have visited before, exclaiming 'Master of Trinity House', the Master in question was writing an article for the benefit of the Brethren: -

'*Mr Winstanley has distinguished himself these past three years with the creation of a Lighthouse 9 miles from shore - the first to be built entirely at sea - a task considered impossible by many fine Architects. Mr Winstanley profits from each ship to the tune of one penny per tonne passing in either direction, and no doubt gains comfort from the knowledge that there have been no recorded shipwrecks since his Light took residence upon that rock.*

My heartfelt congratulations to him,

Respectfully, Samuel Pepys, Master of Trinity House.'

Henry had reached an age where he felt his life worthy of reflection, for after all, he had achieved success with every wonder he had created, so what reason could there possibly be for him not to recreate another one for another King's glory - even if he was the enemy King of France?

Having already created a successful wonder, it is nigh on impossible to create another with the same degree of passion as the first, yet, never one to quit, Henry vowed to struggle on, as much for his own ego as for anything noble.

Meanwhile Jane was performing her usual duties at the turnstile, managing another queue of visitors when she spotted a man with face covered by a blanket, edging through the side door, entering the establishment without paying the fee. Thankfully nobody in the queue had noticed or she would have had some awkward explaining to do, so when she spotted Jack edging through the same side entrance, she let the matter rest, trusting him to perform whatever was necessary.

In his garden studio Henry removed the said blanket, revealing a face in a state of near panic, bending over a pile of half-completed sketches for King Louis' 'Palais de Merveilles'. Rather than revealing anything new, the sketches were copies of what already existed for his own House of Wonders - the windmill, the organ, the pond and the chairs on rails. Therefore, Jack's arrival came as a welcome diversion bringing with him the latest copy of the Gazette.

"Can I bring you anything else, Henry?"

He was met by a single word, delivered bluntly,

"Inspiration."

Jack opened the door revealing the latest group of visitors in the distance, all laughing,

"Your mechanisms obviously still work well enough - can you not just copy them?"

"That is the problem."

"Well, I am not the one to know but I would suggest if it tickles the British, then it will tickle the French."

Henry just grumbled something indecipherable, so Jack thought it best to go and leave him to his grumblings, placing a hot drink next to him with the latest copy of The Gazette alongside.

The newspaper was dated '26th October 1703'.

Henry brushed the paper aside in frustration, and in so doing missed a small article concerning himself lurking on an inside page underneath its headline: 'DESPITE MR SAMUEL PEPYS' ASSERTIONS TO THE CONTRARY, HAS MR WINSTANLEY'S LIGHTHOUSE BECOME NOTHING OTHER THAN AN IRRELEVANT FOLLY?'

The paper remained unread, and Henry remained grumbling.

As his hand tried to inject some enthusiasm into the quill, and his fists continually flexed, Henry's mouth opened wide and emitted a strange sound - at this point I should explain that it was not a sound peculiar to Henry, having been employed on many previous occasions, but mainly in hospitals, on the battlefield or on a deathbed - the unmistakable wail of defeat.

Suddenly, an unworldly roar came from miles away to the West, as if a monster was answering his call from some distant land. As Henry stopped to listen, the monster moved closer as if stalking him, disappearing to the left as if losing his scent. It remained the same for four minutes tempting Henry to believe it had gone. He peered through the windows facing West and saw nothing, save the distant trees gently waving. But the monster soon returned with foreboding, sucking at the rivers whilst all the time moving closer, as if Henry was its prey dividing every line of trees in its path to reach him, getting louder and louder with a great gasping and heaving until all of a sudden it was at him, rattling the panes and banging the doors of this virtual glass cage intent on snatching him.

But Henry's mind was fixed elsewhere - on his baby, alone on that rock at the mercy of the angry Gods - it needed him - he had no choice but be at its side.

Gathering the merest of essentials, Henry forced his way through the collapsing studio doors into a Fury on Earth, gulping at the air with whatever breath he had left, suffocating amidst the sucking and blowing currents of wind. With every thought on reaching his baby, Henry fell into his bushes, gripped onto his tree trunks and leaped over his gates to reach the stable. Old Toby whinnied and grunted his support, but Henry knew he had to be practical, "It would be the death of you, my old friend."

Henry swiftly saddled up Majesty - a young Colt bequeathed by King Charles - the product of a fiery Arabian Stallion mated with a Mare from temperate England. Throwing open the stable stoked Majesty's Arabic genes, until it burst open the stable doors and bolted free to face the Gods in all their fury, ignoring Old Toby's tantrum behind as it whinnied and struggled to break free and serve its master.

Henry was on a mission with face set hard against the elements, ignoring the hailstones aimed at his face battering his cheeks like musket-balls, forcing a path into his mouth, denying him breath whilst peppering his eyeballs, denying him sight. His mind remained fixed upon one thing and one thing only - and nothing was going to stop him, united as he was with the thrusting stallion below him, sharing their pain to blindly leap, lunge and plunge over hedge and across field, following tracks so narrow as to be virtually invisible. It was dangerous enough to gulp at the air for fear of the hail, yet Henry had no choice but open his mouth to gasp at the sight before them now - trees strewn like corpses left to die on the field of a great battle won, lost and gone, the horror of which sucked at what little breath he had left - abandoned houses with broken roofs denying shelter. Henry hid his face wishing to witness no more, sensing Majesty united in despair, losing pace from canter to trot between his thighs, kicking its legs out in front to pull the land back and thereby continue forward.

After what seemed like an eternity, their spirits grew merciful with an easing of the wind, a new taste of salt in the air and veils of rain parting like theatre curtains revealing a backdrop of beauty - a distinct curve ahead, where sea and land met growing clearer by the second as man-made patterns appeared - interweaving lines, that over time made sense as the paths and roads that supplied the means by which townsfolk could survive in their little boxes of stone, which then grew larger into houses, becoming more numerous merging into villages of different shapes and sizes, that eventually grew and combined to form the great Naval City of Plymouth - the gateway to Henry's creation and the reason why he'd braved the elements...

CHAPTER 51
JACK AND JANE TWO DAYS EARLIER

Two days earlier Jane had woken to a stern knock at the door, having fallen into the deepest of sleeps following the night of the terrible storm.

A man's voice she knew well, bore news in a manner she had never heard from him before, "My apologies, Madam, but I think you should come and see this quickly, before he is gone for good."

Jane sat bolt upright, "Gone for good? Who? - whatever do you mean?"

"It is best to come quickly, Madam. Brace yourself for the loss. I have the chair."

That was all Jack had said, but it was enough to have galvanised Jane into opening the doors, barely protected by her nightgown, and face whatever had made Jack so upset.

Jane gasped – in the distance she noticed fallen trees having crushed the life out of Henry's attractions trapped beneath.

"Oh, my Lord, Jack, this is terrible."

"This is not why I brought you here, Madam."

Jack's words shook her, being delivered in such an ominous manner, and she was grateful for his steady hand as he helped her into the wheelchair and steered her outside. Jane dearly wished for some words of comfort from Jack pushing from behind but was granted only his sighs and occasional grunts as he negotiated a safe passage through all the devastation.

"Brace yourself, Madam, there is worse ahead."

"Let me see then, Jack, I will prepare myself for the worst."

Composed in grief, Jane observed the extent of the devastation unfolding as she was guided through the remains of the garden, Jack's footsteps continuing as steady as a pulse behind, resembling a pallbearer bearing the coffin.

Finally, it came into view between the fallen trees – the twisted frame of the studio resembling the skeleton of a huge beast shrouded in shattered glass.

"Is he in there then?" she uttered, surprising both herself and Jack with the lack of emotion, which continued thus - "Where is my husband when practical work needs to be done? Gallivanting on some new adventure, no doubt."

"This is not about Henry, Madam, but I am worried for him. He is facing God's fury miles to the West for something very precious."

"His ego, you mean?"

"Madam, would any man with such an ego put himself at risk of humiliation or death? I have never known him be so passionate about anything."

Those words may not have been intended as hurtful, but they struck a chord with Jane - a truth she would have rather ignored than been made to face so openly, for passion had been absent in their marriage for years.

Jack had more to add:

"I wish to be by his side, as he has been by mine on many occasions."

Jane paused for a moment, knowing Jack knew more, but was holding back,

"Should I gird my loins then, Jack?"

"Over there," came his matter-of-fact reply, sounding all the more curt for being so.

Jane braced herself again as she was wheeled into the garden…

Initially everything was exactly as she had expected – shattered panes, broken chairs, fallen branches, even the rails that had carried so many excited customers into the tree, now hung limp.

But Jack had not finished with her just yet, and to Jane's surprise he continued pushing forward relentlessly, making her face the devastation without compromise until to her relief he came to a halt in an anonymous corner of the garden, away from the public's gaze – a view she did not recognise, having visited it so rarely.

For a few moments, Jane sat quietly fuming, feeling she'd been coerced by Henry's closest friend, but after five more minutes she calmed, and in that state of calm she grew to accept the surrounding destruction, and gradually became aware of something rather peculiar - the stable doors were wide open.

They had been battered off their hinges by some ferocious force, leaving scars on thewooden doors like deep gashes cut into flesh, and there lying on his side was the culprit - Old Toby, his mouth coated in dried froth, as dead as the recently extinct Dodo.

"He must have suffered dreadfully all alone, last night," Jack muttered, genuinely devastated for the beast. "Wherever Henry was heading, Toby wanted to join him. He died being loyal to your Henry."

Jane about-turned 180 degrees facing him head on, and spat the following words with such force they splattered his cheeks:

"What makes you so sure of that, Jack?!"

Jack took a second to wipe his cheeks, "Because Toby truly loved Henry, pure and simple," came the reply with a brevity that halted Jane in her tracks.

As she looked down at the old retainer, memories came flooding back of happy former times when Toby had led them both on happy ventures to the riverbank, the town and the Inn, openly sharing their lives.

Jane came to a realisation – her husband had not changed, he had remained exactly as he was before, searching for truths to solve new problems. No, the change had been in herself.

Tears that had remained stored deep in her cheeks for far too long, now raised up, releasing over her cheeks.

"Please, can you take me to Henry, as fast as you can, Jack!"

"But I have no idea where he is…"

"I think I do - the clue will be somewhere hereabouts, for this is where he would have tormented through the night, until discovering the solution he sought. I can think of no better place to find Henry than here."

The once-sturdy studio had been twisted beyond recognition by the forces unleashed that previous night yet ironically, amongst all the carnage the flimsiest of things had survived - pieces of paper lying in what appeared to be a definite order. As Jane turned the pages the half-finished sketches made clear what had been troubling her husband throughout last night - a windmill, a hall of mirrors, coloured fountains, and a dining room for automatons all in the French style and all dismissed into crumpled balls.

Jack and Jane stared at the evidence and arrived at different conclusions starting with Jack,

"Henry must have attacked his work in frustration – it looks like a French version of your House of Wonders designed for King Louis."

However, Jane was not convinced, for lying at the bottom of all the papers she discovered the answer in the form of The London Gazette, despite being soaked like a towel from the storm, its headline was clearly visible:

'HAS MR WINSTANLEY AND HIS LIGHTHOUSE BECOME AN IRRELEVANCE?'

Jane took a long while deliberating - "Henry has gone to defend his child's honour. How soon can you take me to him?"

"I have been advised by experts it would take at least 6 days, Madam."

Jane's disappointment was both instant and obvious, bringing a tear that prompted Jack to make an unlikely pledge:

"However, we do not need 6 days - we shall do it in 4. If Henry can achieve the impossible, then so can we."

CHAPTER 52
THE CITY OF SHIPS

Henry approached the great Naval City of Plymouth nervous of how he would be greeted, so remained high in his saddle out of the reach of potential attack.

Wading through folk flowing listlessly on either side as if in a state of shock, Henry's overall impression was one of great sadness - the reason for which soon became clear, for nearing the harbour it became obvious who the Gods had chosen to punish that night: this great City of ships had been transformed into a graveyard - its heart having been ripped out, exposing many broken boats, scattered like carcasses, being picked over by opportunist thieves as the bodies of perished fishermen were carted away. Those were special men and those were special boats - not pleasure boats or private boats, but *fishing* boats and those men were no ordinary men but *fisher*-men who had battled the elements to bring a catch to the market.

Henry gazed down upon it all from the saddle, as a detached observer, but that did not mean he did not feel their pain, for he felt their loss as much as that of his own crew.

The first group to acknowledge him did so with disdain, approaching with mouths turned down as much as their spirits - as was the man who led the group with contempt in his eyes and a snarl on his lips,

"This fancy horse is not from anywhere hereabouts. Have you come to gloat?"

"Not at all, sir."

"Don't you bloody well 'sir' me, sir. Who are you and what is your business?"

"My name is Henry."

"We have plenty of Henrys around here - we need no more, so be gone, and take your fancy steed with you."

Other folk approached like a herd of inquisitive cows making a similar noise with their long rolling 'Rrrrr's', befitting the long rolling hills of Devon.

"Greetings," came a younger, friendly voice using the same rolling brogue as the rest, yet delivered without malice, and deserving of Henry's equally polite response for being so,

"My thanks, sir, is there an Inn nearby - I am dry from the ride?"

"If you show willing to tell me your story, then I shall willingly show you an Inn. My name is Ralph Reynolds," (a name making the most of those rolling 'Rrrrs') "I am a news reporter in the employ of 'The Sherborne Mercury'." *

He shared the last detail with a great deal of pride, expecting Henry to be impressed. Yet having already dealt directly with England's foremost newspaper, Henry showed no interest at all…

Permit me to introduce Ralph Reynolds - a nondescript young man of small ambition, content to drift where the currents took him - he did not sing well, nor laugh well, but he did pronounce the most wonderful 'Rrrrrs'.

'Ambition' is a word that referred to his parents more than he, for having failed to reach the top of the social ladder themselves, they left Ralph in charge of finishing the job for them.

But what they had failed to realise was that by bullying Ralph since birth, they had produced a son who had no idea of his own mind and therefore no desire to rock any boat. Consequently, when he landed a job at 'The Sherborne Mercury' his parents had something to brag about, and the newspaper was confident it had found a safe pair of hands to do their bidding.

Over the last 3 years Ralph had created numerous unremarkable articles for them, so consequently when the wealthy local 'Influencers'

came that day to bully the poor fishermen into submission, Ralph initially cared not a hoot...

These self-proclaimed 'Men of Influence' - Plymouth's Bankers and Councillors looked and moved like a shoal of fish - identical in their suits all moving as one, feeling safe in their numbers - so as the fishermen strained their backs to clear the debris, the Influencers made a show of waving their arms about and stabbing their fingers issuing orders as if they were Naval Officers.

However, they all came to an abrupt halt upon spotting this new arrival and his fancy horse, and remained so at a safe distance, waiting for their self-proclaimed leader to speak, which he did beginning with a stammer that grew less and less pronounced as he gained bravado from hearing his own voice,

"Wh- wh are you, and wh-why are you here?"

"Gentlemen, as I have already told this young reporter, my name is Henry and I have come to visit my creation."

Another group was arriving, also dressed as one, but this time their pulled jumpers and patched trousers were the result of necessity - for these were the noble men of the sea displaying the damage, and spreading the odour of their work. Their faces, though, bore the same hostility towards Henry as their counterparts.

That was until one of them asked:

"We know of a 'Henry', one who toiled and brought us the light. He is the only Henry we know of or care for."

"I am he, sir."

Their reaction was both instant and heartfelt,

"Then welcome, and thank you, sir, for keeping us safe."

All the fishermen duly flashed the whites of their teeth, speaking at once like excited children, such was their fervour, and moving as one with arms outstretched to touch their hero.

Henry was terrified, unused to being the centre of such passion from strangers, but here was proof of his worth, proof of what his father had failed to see. And only yards away, Henry recognised a fellow sufferer, another young man unaware of his worth after years of being dismissed – a result of poor parenting.

At that moment, Henry decided to take Ralph under his wing...

"Record everything you hear this day, young Ralph, for they may change their tune by nightfall."

Ralph readied his pencil, to the astonishment of the Influencers,

"Pay no attention to this man, Reynolds - or his fancy yarns. He is not from these parts but a man of the City. This is nought but Metropolitan meddling."

A fisherman interjected with passion, "No, he is saving our lives."

An Influencer from the Bank joined in, "He speaks from greed - wrecks are made by storms from God's own hand. We are all entitled to gather whatever the Good Lord provides."

Another fisherman joined in, "But Henry is protecting us."

"No, he will starve you all!"

"He is trying to keep us safe."

Ralph responded with surprising swiftness, scribbling notes with a fervour he had not felt before, not going unnoticed by his employers who issued a stern warning:

"Be careful, young Reynolds - our readers only need to hear our views, not those of some City Man from hundreds of miles away."

But there was no stopping Ralph, who for the first time in his life felt what it was like to have conviction pouring from his pencil point, despite all the threats being hurled his way,

"Reynolds, stop this minute, or we will need to have words with your father, and The Mercury."

It was as if Ralph was finally seeing the light, listing arguments from both sides in a manner indicative of the best reporters - impartiality:

So, when one voice said - *I no longer feel alone near the rocks..."*

another argued, "He is playing tricks on us all!"

and so, it continued in similar spirit, "Henry has shown what likely killed my brother..."

"He is nought but a Magician!"

"Yet finally, someone is thinking of us...."

As both sides stoked their fury, Henry's rage reached boiling point,

"Are none of you interested in why I have come?"

The crowd were so embroiled in their bickering, nobody paid Henry the least bit attention as he roared, "Tell me - does my creation still stand?"

The noise was bedlam as both groups went for each other tooth and nail, making Henry draw his gun for the first time in his life and discharge it into the air.

Both sides stopped in an instant, and suddenly Henry had their interest,

"Does it still stand?"

One of the Influencers cried out, "It is gone."

"Are you sure?"

"It is gone, I tell you."

Henry pointed to Ralph, "Make a note of this, young Ralph, we need to get to the truth."

"Yes, sir," Ralph replied brandishing his pencil ready for action.

Two of Henry's words had set the young reporter alight - 'Ralph' and 'Truth' - for nobody of importance had ever called him by his first name, and nobody including his employers, had ever told him to report the Truth.

Henry was on fire, "Tell me the Truth – has it gone?"

"It must have gone."

"That is not what I asked - take note of how they avoided the question, young Ralph."

Another of the Influencers spoke out, when he would have been better off, keeping his mouth shut,

"Nothing could have survived that storm."

"'Could have survived' does not mean 'did not survive', eh Ralph?"

Feeling of value, Ralph sprang into action, scribbling notes as if in a fury discarding pencil after pencil that snapped from his rage.

One of the bankers snarled, "Be warned, Reynolds - you only write because we provide the pen."

Feeding off Ralph's anger the fishermen grew angry too, "We all agree with you, Henry."

Henry pressed home his advantage, "Then will one of you take me there?"

"You are mad."

"Aye, that is what many say."

"We are inclined to agree."

"Trinity House said I was mad for wasting my time on such a folly. They refused to fund it, so I funded it myself - take this all down, young Ralph."

Ralph jumped to it, taking dictation, trying to keep up with the following volley of words.

"They said it would not stop the shipwrecks! But now you know it did! And I believe it will continue to do so!"

Such fury from such a tired, ageing man earned the respect of every good man there and of all the women who had turned up to see what their menfolk were about.

"Take him to see his baby, Alf," one of the women cried out above the din to the one man amongst the crowd who was turning away.

His wife continued to do battle, "Go take Mr Henry, Alf, or I will never again let you take me."

Alf trembled, but the crowd rallied behind his wife with such cheering as to diffuse the situation, as much as cold water poured over fighting dogs. Henry was grateful,

"My thanks, Madam, I want to see if my baby is safe - I hope you understand."

"I do, sir - Alf, take him the 'morrow or there will be no supper for you this night an' nothin' else either."

The crowd laughed out their angst, and gradually dispersed whilst Henry dismounted, only now feeling the pain of the ride. Many more of the supportive wives gathered around him, clearing a path through the Influencers.

Alf's wife sidled up to Henry revealing her maternal side,

"It must be a lonely place for you, my dear."

"It is, Madam."

"I do not mean your lighthouse."

"I know."

"Come and drink with us - you must both be parched. Come and hear our song."

"My apologies, Madam, but I really must catch some sleep."

"The song is about you."

"Oh well, if you insist…"

Ego drives every sane man and woman to a certain extent, and Henry proved no exception, or he would have achieved nothing at all, so when he and his ego entered 'The Fisherman's' Arms' that evening it was as a hero and in the best of spirits.

Their song was sung heartily by all, with every one of them knowing the lyrics by heart:

'My father was the keeper of the Eddystone Light,
And he slept with a mermaid one fine night,
From this union there came three -
A porpoise and a porky and the other was me!
Yoho ho! the wind blows free,
Oh, for a life on the rolling sea...' *

The pub continued in full swing with Henry joining in as fully as the rest, when a confident man strode up to Henry in sober mood, not wishing to join the gaiety. He took Henry's hand and held it out straight,

"Keep it there for me to see."

Bemused, Henry held it out as did the man with his.

Although Henry had no idea who this man was, the crowd obviously did as they turned silent and gathered around to watch.

Staring into the man's eyes Henry saw no fear, nor anger or bitterness, just a face evoking the famous phrase concerning 'still waters…'

Henry's arm began to ache but the alpha male in him refused to let it drop - which was stupid, for here was a man at least twenty years his junior, with seemingly nothing to prove. But men being men they held out, until to Henry's relief one of the wives intervened,

"You are both like my dogs - I've a mind to pour a bucket o' cold water o'er the both of ye'!"

Henry decided it was up to him to be the bigger man,

"Greetings, sir, my name is Henry Winstanley. I am the man who built the light and am very pleased to meet you, whoever you are and whatever you may be."

The man replied in a calm manner that befitted him well,

"Mr Winstanley, I believe we are to meet in the morning at 5 of the clock."

"Are we?"

"I shall see you at the Hoe. Do not be late, we have tides to hit."

And with that the enigmatic man aimed for the door, and the crowd duly parted like the Biblical waves to allow him safe passage.

Henry was obviously curious, "Who was that man?"

The wife replied, "That was the only man who can guarantee you a safe passage to the rock, sir. With his knowledge and steady hand, you have the best chance of reaching your baby alive."

A fisherman chipped in, "He is without doubt the best pilot I have ever seen. He knows the rocks better than any other soul and is the only man to have cheated them time and again. There is but one channel through that rocky maze capable of accepting a vessel of moderate size, splitting the rock neatly from North to South - it has adequate depth but is barely wide enough to allow a ship to land provisions. Move too much in either direction and it will split the hull sending you to meet the hungry creatures below. He is a master of his art, sir."

"Aye. I have witnessed him before, sir, persuading the hull to go where he demanded - a push here and a prod there, delicate as you like. It's a joy to see such a big thing alter course from such delicate coaxing -"

"Aye, like me missus an' me."

"Then I both owe him, and all of you, my sincerest thanks."

"And also, your life."

"Indeed. What should I call him?"

"By his name, sir – Bound, James Bound - a much sought-after pilot, for none but he can navigate a safe passage through those rocks. Some have believed they could -"

"Aye, but they are the ones providing the Kraken with its supper."

Henry's ears pricked up, "The Kraken? But that is nought but a myth, surely?"

"If it helps you to sleep better in believing so, then yes."

"So, I take it you believe in the Kraken then?"

"I believe in Men of the Sea, sir. Others shift their beliefs too readily, but sailors stay firm and true. If they believe in it then so do I."

Henry glanced at Ralph, hoping he had written it all down - the look he received indicated that he had.

That night Henry went to bed exhausted, yet he could not relax until Majesty's care was assured, as well as a promise from Ralph Reynolds to meet him at the Hoe at 5am then sail to the rock to see his baby again, whether it be still standing upon the rock or lying broken beneath the waves…

CHAPTER 53
HENRY'S BABY

5 *a.m. came too swiftly for young Ralph Reynolds - after an evening of sea shanties and rum he had collapsed into bed and now, five hours later, was staring at the clock, dreading the secret he would have to divulge to the others....*

The wind had slackened during the night - not completely, but leaving a short window of opportunity, if one was daring enough to take it. The fishermen recommended caution, which relieved Ralph considerably, but he was about to learn that if Henry had his mind set upon something he would not back down, and if the odds were stacked against him, then he would be all the more determined to carry it through.

"Should we not put our trust in the fishermen and delay, sir?"

"We should not. If I acted upon every piece of caution, I would never have achieved anything."

Henry set off at a pace compelling Ralph to follow, despite his dread and the dragging of his feet.

The quay grew closer with every step, and with every step Ralph felt his chances of backing down diminishing. A sudden burst of laughter from above jolted him - the gulls were playing on the wind mocking him as if they knew better - and Ralph considered that probably they did - for they will have seen what was coming over the horizon. To Ralph's dismay, Henry refused to slacken the pace, even speeding somewhat to see his baby, compelling the fishermen to follow suit and Ralph to do likewise - each step committing him more to the cause. The only member of the group refusing to comply was James Bound, continuing his steady pace, as sure of himself as any man would be, having been congratulated so many times to make him feel invincible.

Ralph's eyes widened as they passed the rows of rowing boats along the quay checking them over as he passed - the oars for breakages, the rowlocks for the state of their attachments - all of which looked fine to him whilst realising if they were not, he would be none the wiser anyway for he knew nothing of boats, wisely putting his faith in those that do. Hearing the boats frightened Ralph with their banging and clattering, as they bumped into one another bucking like young foals eager to be released.

Ralph felt likewise, "This is madness, sir, these boats could splinter in powerful waves."

The fishermen burst out laughing joining the gulls, which Ralph reasoned to be bravado concealing their own fears, for their warnings had been heartfelt. Finally, Henry came to the rescue with knowledge and wisdom,

"We are not rowing out this day."

To which they were all relieved, none more so than Ralph, who risked being thought stupid again: "I thought not, sir - all those stories about you rowing 18 miles each day for just four hours' work were ridiculous - and at your age too."

"Aye, they were ridiculous," replied one of the older fishermen, "For sometimes it was so rough, we could not land and had to turn straight back."

Another added from behind, "I'm even older than Henry, so be careful what you say, young lad, we are getting it easy not having to row today because of you."

Soon Ralph saw what he meant, for a sailing ship was moored further ahead waiting to fill its sails, bringing with it a degree of comfort, "Thank you for the ship, sir, but are you still certain you want to go ahead this day? Many ships left harbour yesterday only to turn back. I am told all ports are crammed full."

"I am certain."

"But we could still turn back, sir, before it is too late."

"You should never turn back on a commitment, Ralph."

"Have you not ever changed your mind, sir?"

"I have."

"There, you see, let us turn ba-" but Henry interrupted,

"A long time ago my father told me, I should stop my innovations, change tack and join the Church. I was about to take his advice, then I thought about it, changed my mind and my life has been made all the richer because of it. Now follow me, and remember, do not show weakness - commit and do not turn back."

Henry led the way boarding the sailing ship, admitting some relief himself at not having to row in the increasing swell, "Keep your eyes open and your paper dry, young Ralph. I am relying on you to convey the truth of our journey this day, so your readers can better understand."

"Understand, sir? - understand what?"

"Why people like me do what we do, so that people like you can write about people like me."

Looking over the side Ralph spotted two men rushing up to free the ship of its moorings. To Ralph's horror, the ship rocked hard as if relieved to be free then swung about heading due South to face the lines of waves ahead full-on resembling opposing troops approaching on the field of battle - their beating drums being provided by the waves hitting the hull.

"I have a confession," admitted Ralph at last, his voice trembling with the fright of it, making his words indecipherable.

"You have a what?"

"I have to confess something most terrible, sir. I am prepared to leap overboard if you command me."

"It is too late for that, Ralph. Look over the side."

Ralph took his advice and staggered across the tilting deck to look over the side, collapsing to his knees on the way. Henry laughed,

"Do you really want to jump into that? Stay here, I need your report."

"I am sick, sir, I am so sorry, sir, please forgive me, but I suffer from sea sickness. You all must think me stupid. My apologies."

But to Ralph's surprise, Henry laughed heartily, "I was as sick as a dog before meeting King Louis," he announced recollecting vomiting in the French privateer's hold.

"And I am sick every time I go to sea," admitted one of the fishermen generously.

"Look at the horizon - that helps me," advised another.

"I have just been sick," announced another, attracting laughter from all the others.

James Bound was a man of few words, so when he did speak, it was worth listening to:

"Be quiet."

The ship continued heading South-South-West almost directly into the prevailing wind, with no other ships for company, not even the gulls who, being an intelligent species, had already fled for calmer climes.

Ralph took the opportunity to try writing his notes but soon found the spray and the rocking rendered the task impossible, not helped by the result of his ham and egg breakfast sliding to and fro across the page.

"Good boy, that should be a lesson learned," smiled Henry kindly, "It's out now, boy - and once out cannot come out again."

Ralph was mesmerised by this father figure staring straight ahead for a sight of the baby he loved, constantly denied by the ship as it rose and fell banging over the waves - yet despite the wind and the spray attacking his eyes, Ralph noticed this man managed to retain the youthful spark that singled out Mr Henry Winstanley Gent. from others of his age, leading Ralph to hope that he too would be as driven at 59.

"Can you see it yet, sir?"

"Not yet, it must be hidden between the waves," Henry replied hoping it was not actually *beneath* the waves. He was growing fond of young Ralph, recognising some of his own naïve enthusiasm at Ralph's age, and couldn't help but wish he had been blessed with a son of his own,

"Look ahead, Ralph, it is all out there for you if you are willing to take it. Do not hold back."

"I will try not to, sir," came the reply but it was as if he had not replied at all, for Henry remained oblivious staring straight ahead, fixed in a daze.

Ralph fell silent, allowing the old man time alone with his thoughts, so was surprised when Henry suddenly sparked up with: "Does your father guide you well, young Ralph?"

"He does not guide me at all, sir."

"Good, then guide yourself for nobody understands you better than yourself."

"I shall do, sir."

Ralph thought he detected something in Henry's eyes – a paternal acknowledgement, but he couldn't be sure. Then -

"Good Lord!"

Henry suddenly reached up leaning forward. Ralph stretched up too but saw nothing save the horizon rocking to the left and right, now deserted of gulls, prompting some of the fishermen to comment, "Even the gulls have deserted us. Aye, never before have I known such a thing. It does not bode well…"

Henry surprised them all suddenly laughing and stamping his feet like a two-year-old,

"Hello, hello, my dear boy."

Ralph jumped forward grabbing at the rail, seeing only the horizon shooting up, then down, revealing sky then surf, then sky then surf, then sky then surf, then a distant fleeing gull, then surf, then sky with something other than a gull, then surf, then there it was - the tip of metal that promised the weather-vane beneath, then surf, then what was definitely the weather vane again, good and solid and true.

"Bless you child, oh, bless you," spoke Henry, barely above a whisper, "You look good, I must say you are looking very good - a little more worn perhaps, but never mind."

Ralph tried to make notes, but the paper disintegrated between his fingers, it being so wet. At first, he wanted to cuss but remained silent in respect of Henry's private moment…anyway, he would have no need of noting this down, it being etched into his mind forever.

Each rise and fall of the bow revealed more of the lighthouse, until it was fully revealed, strong, shocking and glorious, standing bold upon the rock - 115 foot of magnificence beyond Ralph's writing skills to do justice - and a sight he would be seeing again every single night, from that day forth at the closing of his eyes. And he admitted to feeling jealous of this parent's love for his child, having felt so little from his own.

Henry was overcome, eager to savour touching it once more as he laughed,

"They said you were gone, but I knew you were here – I could still feel you in my heart. Take me to him, James, I want to touch him."

As James Bound prepared the landing boat, Ralph noticed a flash of doubt cross that normally stoic face, acknowledging the degree of responsibility now relying solely upon his celebrated skill. Unusually, he began by quietly taking his time checking to the left and right, weighing up the balance between the solid rocks and the bouncing ship, for he knew he had but one chance in this narrow channel in these conditions. Checking, then checking again to the left and right, suddenly, his eyes lit up and his bellowing began -

"One pole to the left and one to the right. Hold steady till my command - steady, steady, steady, I say…only push when I command."

These fishermen were well trained in the art of seamanship and knew the importance of responding swiftly to any command. Grabbing a pole each, they took position along either side of the deck easing their poles out, like oars on an ancient Trireme, and like the Roman slaves of old, they fell silent awaiting the command.

Bound knew this could rank as his greatest ever achievement - he knew it and took ownership of it, and as the ship twisted in that narrowest of gaps, Bound saw his moment and pounced like a beast onto its prey, dodging from left to right bellowing commands,

"Push hard to the left! Hold tight on the right!"

Fortunately, these fishermen had put their complete trust in him, knowing there could be no hesitation so responded in kind, pushing and prodding their poles against the rocks following his commands keeping the ship cantered despite its bobbing and twisting, inches from jagged rock on either side. Ralph looked down from the deck above, absolutely terrified knowing many had witnessed what he was seeing now, before dropping over the side to feel their bones being mashed between rock and hull like Devil's teeth, before sinking into its belly below.

Bound's face, normally so still was now so alive, eyes dodging to the left and right, nostrils flaring, yet for what must have been the first time in his life, actually revealing self-doubt, proving even the sternest of faces can contort when required, and the most masterful of men can betray doubt, both surprising and frightening the crew,

"I have never seen him move so fast."

"I have never seen him so anxious."

Ralph got caught up in the moment, "Anxious for his reputation?"

"No - for his life."

"For all of our lives."

Then Bound saw his moment, "NOW Lower the boat…NOW I tell you - FAST!"

The crew grabbed at the ropes, kicked the landing boat overboard, and tugged at the ropes to keep it from tipping over.

"ALL IN NOW - FAST!"

Ralph held back, terrified until Henry took the lead, and with a cry to inspire his men, dropped like a stone into the tiny boat, tumbling to his knees, making the crew gasp, and where he remained for a few seconds regaining his breath, until he forced himself upright earning a loud cheer, inspiring Ralph to follow suit landing heavily on his butt. The tiny boat rocked - two fishermen jumped in rocking it some more, until four men were in that boat and the fishermen showed their experience taking control, easing it to the rock's edge. Bound eased the ship away from one danger into another, once its sails, suddenly plump with wind hurled them all into a mad world of waves.

An old man's cry rang out, blown by lungs strong enough to be heard above the frenzy. Looking up, Ralph saw two piercing eyes staring down amidst a face smothered in curls like kelp when a hole appeared with teeth emitting a voice both strong and clear, "Greetings, Henry!"

Henry shouted back as loud as he could manage, "Greetings, Lucian!"

Lucian's voice proved it could penetrate as strong as Henry's candles above his head,

"Grab the swing."

Ralph was amazed at how fast the lighthouse came to life swinging out its arm with ropes attached like bracelets reaching for the swing seat. Henry climbed on board with difficulty as waves drenched his back, and for the first time in a long time, he ascended his creation, marvelling at his construction, as if seeing it for the first time, checking the new iron handles, and noting how well the new brass plates were enduring. Satisfied, he reached the door and shook his Keeper Lucian by the hand - a mighty handshake for a man of four score years and ten.

"Welcome aboard, Henry - you had best get inside and rest."

Ralph had watched it all from the rowing boat, dreading his turn but looking forward to some shelter from the freezing waves pounding his back, dragging at his trousers then returning to pound his back again. As soon as the seat dropped in front of him, Ralph went into action grabbing at the ropes and dragging himself on board with eyes shut tight, before being hoisted upwards in a series of jolts as the spray drenched him from the exploding waves. Finally daring to open his eyes Ralph caught sight of the huge 'WINSTANLEY. GENT' signage as he drew alongside followed by 'OUR LORD. ANNO DOM. 1699' - making him smile at Henry's audacity, advertising himself on an equal footing with God.

An old hand suddenly gripped his arm, and pulled him up into a world of calm and relative quiet…

Lucian looked superhuman with his wrists of steel, and eyes of sapphire that still shone as bright as when he was 15 - 4 score years ago. He was not alone in that place, having a youth as his assistant - the diligent Tom.

Ralph was overwhelmed – the place was akin to a 12-foot-high stateroom adorned with oil paintings on the walls, plus shelves lined with bowls of lead crystal recently smuggled from Venice,

"Good heavens, Henry, to all extents and purpose this is a state room 130 feet above the waves."

"Exactly right - and exactly how I want you to describe it in your report."

"My report?" Ralph stammered, reality kicking in, ending this dream with a jolt, "Of course, sir."

"You must list all the details then leave; I want that report to spread astonishment and awe. Now come with me -"

Ralph was given a swift tour of the place - the fully furnished bedrooms, the bespoke kitchen with its large chimney, oven, dresser, tables and the storerooms, Henry hurried him along, "Lucian will supply you with paper - write an honest account of this place and make it fast – quickly now, before you go."

"Before I go? But, sir, I want to see more."

"The storm will soon be upon us – leave now, for your own sake, my boy."

"What about you, sir? Most are saying this place is not safe."

"'Most' are wrong. I built it. I know every stone, every timber and every damn joint. It has withstood both wind and waves many times, and will do so again. Now go and write your report."

Lucian roared loud enough to burst the ears, "Aye, that sky looks as hellish as I have ever seen. Whatever lies out there will be blown here within the next few minutes."

"Please let me stay, sir."

"We have eighteen minutes left – go, now."

"You care more for me than my father."

"I care for what you write. Now be off with you."

"But what should I tell them, sir?"

Henry was at bursting point, "Tell them this is the safest place in the whole of England. Give them this headline: - 'HENRY WINSTANLEY WISHES TO BE IN HIS LIGHTHOUSE FOR THE GREATEST STORM THAT EVER WAS'.' Now go and be done with you!"

Lucian dragged Ralph across to the window, then man-handled him to Tom who pushed him onto the seat as Ralph called out,

"I shall return tomorrow with my report."

Henry looked down from the window, witnessing Ralph being pushed into the rowing boat as the ship drew alongside, ducking and diving in that narrow channel between the rocks, flirting with danger.

"My goodness," said Lucian in a strange tone bringing Tom alongside in an instant -

"Look at this," Lucian called out, tapping the barometers in disbelief,

"Mercury is at 28.47, I have never seen it drop so low."

Henry looked down through the window, catching Ralph being dragged onto the ship, James Bound throwing his arms about yelling orders, two mighty waves racing to meet each other colliding in an explosion of surf, the ship's sails bang from being suddenly filled, the swing seat directly below twisting like a trapped eel, then smashing into the lighthouse splitting in two - the only chance of escape splintering into submission. The ship's sails full to bursting, were suddenly caught by the wind, jettisoning it away with James at the stern looking mightily relieved to be so.

"20 minutes you said?" Henry asked of Lucian.

"Pardon?"

"Before it arrives?"

"Aye."

"Well, that makes it two minutes from now."

A creak in the structure drew Henry back to the window,

"Good, it is bending like a tree in the wind. Just as I proposed."

The sky was black. No stars, no moon, just black.

Hollow black.

"Tonight is a new moon, is it not?"

The ancient keeper replied, "Aye, it is,"

A huge thud against the window jolted Henry back against the wall - a gull was staring at him straight in the eye, spread-eagled with snapped wings - a peppering of blood decorating the glass.

Then another bang, "Oh, my Lord."

A haddock had joined the gull slapped against the glass, pinned there eyeing the gull with gulping mouth, the hunter and the hunted eyeing each other just out of reach.

"Madness…"

CHAPTER 54
THE WRATH OF GOD

*D**ear Reader, we find ourselves two days earlier on 24**th* *November 1703 at Poundbury, 170 miles West of Littlebury, as a unique occurrence in Great Britain's History was about to be unleashed...*

Crackles of light split open heaving clouds like corsets releasing torrents of pent-up rain, as blasts of thunder cursed whoever was foolish enough to be out on a such a day.

By that reckoning, both Jack and Jane were insane fools, yet in their defence - they were on a mission of mercy:

"This is madness, Madam; we should find an Inn to rest immediately!"

"We have not come all this way to rest, Jack – now carry on."

The two young horses had obediently pulled their carriage through slush and mud for over 200 miles in the early hours of that fateful day, heading South West through London squealing from all the manner of destruction crashing about them, whilst being denied the sight of chimney stacks collapsing beyond the view of their blinders, or roof tiles spinning overhead like sycamore seeds slicing into opposite walls, as folk dived for cover into any house they could find, fearing an earthquake was upon them. Yet, despite the barrage of rain ripping at his face, Jack resisted panic in order to maintain a medium pace, wisely resisting the temptation to race and wear out the horses.

As they crossed various County lines the wind remained their constant enemy, from the blasting rain through Surrey, and the hurling hail through Winchester, attacking his eyes to such a degree it forced him to continue the journey with head held down, hearing his gallant horses struggling for breath through nostrils clogged with sleet.

They battled on until the coast loomed near Exeter, the storm abating sufficiently for them to witness an apocalypse - 300-400 boats of various sizes piled on top of one another, many overturned, looking like beetles in a mating frenzy, plus even more having been launched inland by a tide above any previous watermark. Determined to somehow reach Henry, Jane and Jack continued Westward, witnessing more madness along every mile, around every bend, over every mound - cows falling into ditches, floating bales of wool attached to the drowned ewes beneath, windmills whose sails were designed to catch the wind now overcome, spinning so fast they burst into flame like giant Catherine Wheels - jackets, coats, and shirts blowing everywhere, some with their owners still inside, being hurled about with their limbs poking out waving at nothing in particular - people reduced to lumps of meat no longer bearing souls.

Eventually, and thankfully, Jack and Jane reached the outskirts of Plymouth, beyond exhaustion, finally coming to a halt at Bovisand where they could go no further, perched on top of the cliffs.

"I cannot get you any closer to Henry than this, Madam."

"Point to Henry, so I may get a sense of what he is going through."

Jack pointed over the crashing waves to where the horizon should be,

"He is 9 miles out there, Madam, God preserve him."

Jane peered out but saw only thick cloud above crashing waves, with no discernible joint between, "Oh, Henry."

Dear Reader, at this point most of you would have the mind that it was folly for Jack and Jane to have even tried, but Jack had defied the Gods in order to try and fulfil a promise to unite Jane with her Henry, and in their anger the Gods had taken revenge – lifting the carriage and hurling it through the air, casting all the carefully packed provisions to the wind including the brave young horses which were hurled over the cliff edge, still shackled and screeching their dismay, leaving Jack and Jane alone to face whatever the Gods had in store for them…

However, the Gods are known to be bountiful if the mood takes them, and this day they chose to conjure a miracle - for by hurling waves against the cliff, the tremors had opened sinkholes on the surface, providing an opportunity not to be missed,

"Hold onto me, Jane."

"I have never held any other man but Henry."

"Then you had best start now - jump!"

Jack pulled Jane close to his chest, protecting her from the tunnel walls and as such they disappeared down the sinkhole holding onto one another - Jack, a protective soul remaining true to his nature - Jane, a loyal wife, remaining true to her vow.

CHAPTER 55
THE END OF THE END

Henry had noted the concern on James Bound's face as Ralph boarded his ship, the fear as the mighty wind span him around, and the flash of relief as it hurled him north for the safety of England.

Lucian approached, "Move away from the glass, Henry, lest it break."

"Nonsense, I built it to last." came the reply of a man, maybe too confident, for his own good.

Henry watched as two mighty waves collided too close for comfort, heard the explosive impact plus the creaks and groans of the timbers like an old man's bones, but it was the significance of the next sound that froze his blood to the core - the high-pitched squealing from above, piercing his ears,

"Lucian - save the Light!"

As his eardrums tried to cope with all the squealing, Henry raced up the narrow staircase adding his gasps to a concerto of booming waves; delicate, clinking decanters, the percussion of Lucian's boots closing in below, the falsetto shrieks from Tom, as Henry's artwork fell from the walls accompanying the cymbal crashing of Venetian decanters, all echoing up the hollow tower in orchestrated madness reaching Henry's ears at the very top, now discovering the source of all the squealing - the metal weather vane forced to spin with such fury it fired sparks, severed its stem and toppled, spinning into the turbulence below.

Lucian's eyes, which had witnessed so many violent storms in the past were now flooding with tears, but Henry was already beyond that, peering through the cracked glass at a movement out to sea heaving like some huge creature rolling below the surface - something he had heard tell of but never believed would see: - something was approaching, a thing so powerful it eased the wind and calmed the sea. Henry gasped, forming a cloud of breath hiding what he was desperate to see – with a wipe it disappeared, leaving a clear vision he could not deny – it was as if the Kraken was raising its skirt from East to West, dragging up the sea and all of its creatures in a great wall that raced for him with a mighty hiss, and smashed the glass devouring all Light…

EPILOGUE

Henry's wish was granted - on 26th November 1703 'The Greatest Storm that Ever Was' killed up to 15 thousand souls, and swept away Henry's lighthouse with him inside, never to be seen again.

It is still ranked as Britain's deadliest storm, now believed to have been caused by the edge of a tornado blown across from North America.

No shipwrecks were recorded for the five years Henry's 'Shed on a Rock' stood guard at the Eddystone.

Within two weeks of its disappearance, a Merchant Ship struck the Rock with the total loss of its cargo, and all 57 crew.

Ships continued coming to grief on the Eddystone at the rate of 22 a year, finally prompting Trinity House to pay for a lighthouse on the Rocks ever since, earning 'Light Duties' of one penny per tonne from passing ships, despite refusing to invest a penny in Henry's 'mad' endeavour. Not bad for a man they called a 'Mere Showman'.

By a strange coincidence, at the same moment Henry's model collapsed at the House of Wonders.

Henry's Waterworks remained a popular tourist attraction on London's Piccadilly for over 30 years, returning a profit for his wife who entered a relationship with another creative man – the French wrought iron worker who fashioned her gates at Henry's 'House of Wonders' and the gates of Christopher Wren at St Paul's Cathedral –

M. Jean Tijou.

Henry was 52 when he started to construct his lighthouse, rowing 9 miles each way with his small crew for every day the weather allowed.

Henry was 59 when he died.

A few days later, a ship carrying Tobacco from Virginia to Plymouth was wrecked on the rocks with the loss of all cargo and crew.

Henry's lighthouse surpassed Alexander the Great's Pharos by being built entirely at sea.

Alexander's Pharos is judged to be 'One of the Seven Wonders of the Ancient World'…

Henry Winstanley Gent. - 31-3-1644 - 26-11-1703

THE END

HENRY'S SELF PORTRAIT with the kind permission of Simon Hilton-Smith of the Saffron Walden Museum (credit: SAFWM:1880.2), as featured in CHAPTER 25 – 'TRANSFORMATION'. This is the only representation of what Henry actually looked like. If his engravings of Audley Manor are anything to judge by then this picture is sure to be an accurate likeness.

AUDLEY END MANOR as it looks in 2025 – a third of the original size when young Henry worked there as a porter and was so attractive to King Charles with its proximity to Newmarket Racecourse that He bought it.

CHARLES 11 – this makes me think of the scene in the Mayfair Club CHAPTER 32 – 'LONDON' with the Merry Monarch enjoying a glass of the newly introduced 'Champagne' in the company of pretty ladies. (credit : Alamy Pictures – 'The Merry Monarch at home')

KING LOUIS XIV – leader of France from a child of Four until his death at Seventy-Seven, in which time he had made France the greatest power in Europe. He enjoyed a huge ego with an equally large appetite for music, women, dance and hunting. It was Louis who decided Henry's fate after he was thrown at his feet naked and wrapped in chains as described in CHAPTER 47 'REPETITION'. (credit: Alamy Pictures)

St MARY'S CHURCH, SAFFRON WALDEN with the kind permission of Revd. Jeremy Trew consisting of a print of a copperplate engraving made in 1784. Unfortunately, there are no visual references of Henry's ingenious planetary mechanism. (credit: Revd Jeremy Trew Saffron Rector)

Winstanley's Wonders

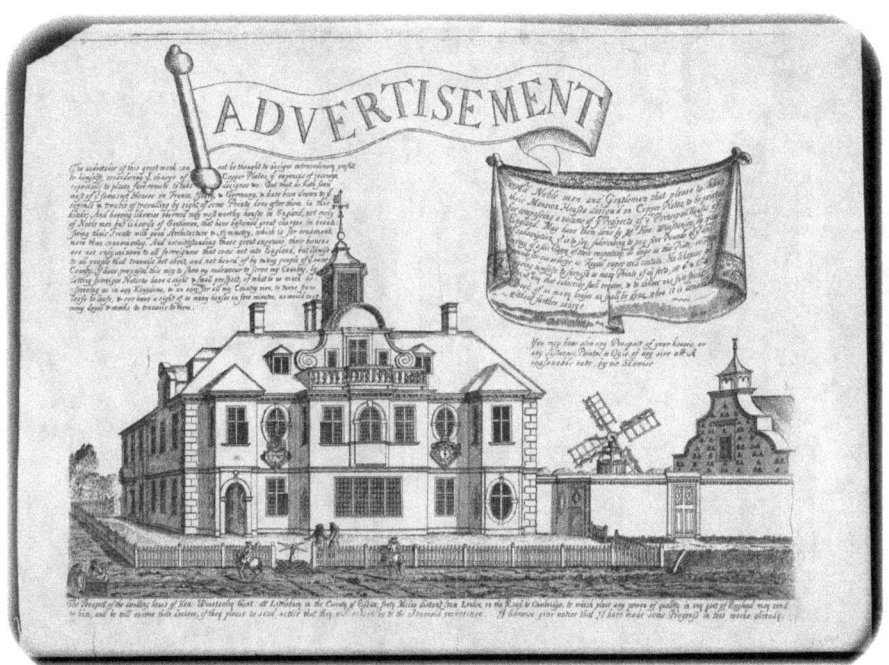

HENRY'S HOUSE OF WONDERS in Littlebury as described in CHAPTER 30 'THE GIFT OF INDEPENDENCE' showing the turnstile where Jane used to take admissions plus the Windmill that powered all of Henry's devices. (credit: Look and Learn)

HENRY'S GREATEST WONDER – a light on the deadly Eddystone Rocks solely paid for by Henry who even rowed the 9 miles out and back with a small crew almost every day to build it despite foul weather and being in his fifties. It was a great success, stopping shipwrecks immediately prompting doubters to lay the success at the door of improved navigation. Enraged, Henry stated his wish to be inside it during 'The Greatest Storm That Ever Was' to prove its worth. (credit: Look and Learn)

HENRY LOOKS OUT FROM THE WINDOW - Henry's wish came true in what is still regarded as the worst storm ever to have hit the British Isles on November 26th 1703, hurling boats, men, women, their kids, dogs, cattle and horses miles inland whilst forcing windmills to spin so fast they caught fire like flaming Catherine-Wheels. (credit: Look and Learn)

DESTROYED – Henry's work was completely destroyed leaving no trace of the lighthouse or the amazing man who had created it. Two days later the Rock claimed its next victim with the loss of all cargo and crew.

If you look closely you can see 5 figures being cast into the maelstrom.

Henry had even surpassed Alexander the Great's Pharos by being built entirely at sea, becoming in effect the Eighth Wonder of the World.
(credit: Look and Learn)

FOOTNOTES

Page:

49 – 'Italian curved points' – shoes with elongated toes became popular mainly with women of social standing from the 17th Century.

76 – 'nincompoop' was first used in the 1670's from the Latin 'non compos mentis' meaning mentally incompetent.

63 – In August 1672 the Dutch Prime Minister Johan de Witt, along with his brother, was set upon by a militant mob, shot and mutilated then having their livers cut out and eaten.

126 – 'upstart four-wheeled tortoises' – the view expressed by some for the newly introduced two-seater horse and covered carriages appearing on the streets

202 – 'riding in the ring' – from 'running at the ring', an equestrian contest enjoyed in the Royal courts of Europe

243 – 'with bated breath and whispering humbleness' – a quote from Shylock in 'The Merchant of Venice.'

307 – 'The Sherborne Mercury' was Dorset's first newspaper that began publishing from February 1737

313 – this is an accurate rendition of the song according to 'genius.com'

269 – 'corrupt doting rogues' – the elders of Trinity House as described by Pepys in his diary

ACKNOWLEDGEMENTS

- 'HENRY WINSTANLEY AND THE EDDYSTONE LIGHTHOUSE' – Adam Hart-Davis
- The late TV PRODUCER MIKE WEARING, one of the best Producers I have ever worked with, who suggested I include a Workshop for Henry, akin to Q's lab in the Bond Films
- NEIL JONES – Trinity House Communications Manager
- The MINISTRY of HISTORY
- 'TRACERY TALES'
- Miniatures of Henry's creation variously advertised in California
- OXFORD DICTIONARY of NATIONAL BIOGRAPHY
- thefreelibrary.com
- Simon Hilton-Smith of Saffron Walden Museum for granting me unrestricted permission to use Henry's self-portrait
- Reverand Jeremy Trew who granted me permission to use the historical rendition of his Church showing Henry's lantern attached
- D Norman of 'Blender Artists' for allowing me to reference his artwork of Henry's Lighthouse as part of my front cover
- genius.com for confirming the Eddystone Light lyrics of the sea shanty